THE NEW KILLER DISEASES

elinor levy

AND

mark fischetti

THE NEW KILLER DISEASES

HOW THE ALARMING EVOLUTION OF GERMS THREATENS US ALL

Published by Three Rivers Press, New York, New York.
Member of the Crown Publishing Group, a division of Random House, Inc.
www.crownpublishing.com

Three Rivers Press and the Tugboat design are registered trademarks of Random House, Inc.

Originally published in hardcover in slightly different form by Crown Publishers, a division of Random House, Inc., in 2003.

Printed in the United States of America

DESIGN BY ELINA D. NUDELMAN

Library of Congress Cataloging-in-Publication Data
Levy, Elinor.
The new killer diseases : how the alarming evolution of mutant germs threatens us all / Elinor Levy and Mark Fischetti.
1. Drug resistance in microorganisms—Popular works.
2. Communicable diseases—Popular works. I. Fischetti, Mark. II. Title.
QR177.L475 2003
616'.01—dc21

2002155925

ISBN 1-4000-5275-0

10 9 8 7 6 5 4 3 2

First Paperback Edition

CONTENTS

THE NEW KILLER DISEASES

INTRODUCTION

SARS, THE NEWEST KILLER

This book was going to press when the news hit that a mystery illness, SARS, had erupted out of China and was rapidly spreading across the world. Dramatic events unfolded each day, events that demonstrated the urgent dangers of new killer diseases that we write about here.

The story of SARS (severe acute respiratory syndrome)—why the virus spread so rapidly, and how the international community responded—is a powerful, cautionary tale about the thorny challenges we must grapple with as we confront a growing number of infectious microbes that are evolving at an alarming rate. In the chapters that follow, we explore these pressing issues, introducing all the vital information the public needs to know about how best to protect ourselves against these diseases, as well as measures the medical community and our government should be taking to fight them. SARS has unfortunately established a significant foothold around the world, and we must expect that it will continue to spread—though less rapidly—and also to evolve, just as West Nile virus, Lyme disease, and other recently emerged scourges have done. SARS is one of the fastest spreading and most virulent new diseases the world had seen in some time. Yet we might well face an even more deadly foe before long, and we must learn the lessons the SARS saga holds for us.

The good news witnessed during the SARS outbreak is that the world community has made great progress in its ability to track and combat emerging diseases. But the outbreak also brought into stark relief a number of weaknesses in our defenses and a set of tough questions we must address. When a new disease has broken out, can we depend on all nations to share information readily and to take the necessary steps to limit the spread? What are the most effective means of containing a new disease, and what are the acceptable limits of government control? In the case of SARS, did the World Health Organization and national health authorities act fast enough to warn people of the danger, or did they go too far in some countries, provoking unnecessary panic? What social costs are we willing to pay in order to respond with the vigilance required? Do we need more disease detection at airports and other points of entry, or tougher quarantine laws? What will the ripple effects on the world economy be? And most fundamentally, how can we best determine that a new disease has appeared, and why it spreads so much more rapidly in one locale versus another?

The World Health Organization (WHO) sounded the SARS alarm on March 15, 2003, declaring the disease a worldwide health threat because it was spreading so far, so fast. In less than seven days after the first case had been identified in Hong Kong, dozens of people, many of them health care workers, had come down with it there, and patients had also been diagnosed with SARS in Hanoi, Vietnam, and Toronto, Canada. The fact that the disease had made it all the way to North America so fast alerted the WHO that this outbreak might well become a global epidemic.

The first symptoms of the disease were fever and dry cough, and in some people muscle aches, sore throat, and diarrhea. These symptoms mimic those from influenza or the common cold, so at first patients were unaware of how serious their illness was, and many didn't go to see a doctor. However, SARS quickly progressed to pneumonia in some patients, or caused such difficulty in breathing that doctors had to hook some up to mechanical ventilators. Those who succumbed were killed when, as in other pneumonias, the heart or other organs failed due to the stress of not getting sufficient oxygen.

More than any other recent disease, SARS traveled like wildfire because of "superspreaders," people infected with a disease who pass it on to an inordinate number of others. In Singapore, for example, one young woman who checked into a hospital after returning from Hong Kong infected twenty nurses and patients, while two other young women who had also come back did not pass on the germ to anyone. Evidence for superspreaders is rare in history. Occasional AIDS patients and TB carriers may have fit this label, and certain people exposed to any ailment such as flu are more contagious than others, but the SARS superspreaders stood out as flagrant examples. Ironically, it was the unusual spread of SARS from a single patient to many of that same person's doctors and nurses that first alerted the medical community this disease was different from other pneumonias. Indeed, it was the stunning specter of large numbers of hospital workers falling gravely ill in several of the treatment centers where patients were being examined that prompted the WHO to issue a global warning.

As of early May it was still unclear if the superspreaders were somehow different from other patients, or if they were just at the bad end of a spectrum of infectivity. They are a topic of great interest, but little is known yet. Do they have a differently mutated form of the virus? Are they co-infected with another virus that amplifies their contagiousness? Or are they just more susceptible to the virus because of a weak immune response, which allows the disease to proliferate that much more wildly in their bodies, so that their breath is literally saturated with virus particles? Unlike Hong Kong and Toronto, the United States was lucky that a superspreader had not traveled to the country, which is the reason so many fewer cases occurred here.

Although the news of SARS erupted on March 15, as we now know, the disease had already been coursing through southeastern China for three months but that was kept quiet by the Chinese government. Officials at the WHO who worked inside China apparently received some word about the problem, but either they were not given enough information or they reacted slowly to the snippets they did receive. The first glimmer that a truly dangerous situation was developing occurred in November 2002, when WHO officials attending a regular influenza meeting in China heard that several people in the southeastern part of the country had died from a severe pneumonia. These deaths raised a red flag, but the local analyses of virus samples from the patients showed nothing unusual about the virus.

Then in December, Hong Kong newspapers reported the deaths of wild ducks and geese at a park within the Sha Tin racecourse in Hong Kong. After inspectors discovered that the fowl had been infected with a flu virus, they closed the park and destroyed all the remaining birds. Officials in Hong Kong are all too familiar with the story of a 1997 outbreak of H5N1 avian influenza that killed a 6-year-old boy and prompted the slaughter of millions of chickens due to fears that an epidemic might be kicked off. When flu viruses jump from an animal to a human, they can be extraordinarily virulent. That's why officials in Hong Kong are always on alert, especially considering the annual preponderance of flu outbreaks in the region. They felt confident that by their speedy action in this case, they had addressed the threat.

By early February 2003, however, rumors in China were spreading about more unusual human pneumonia cases, occurring in the southeastern province of Guangdong, and those rumors reached regional WHO officials. Then a local doctor told them that a 33-year-old Hong Kong man had died of an aggressive flu-like illness, and that his 8-year-old daughter had also died of pneumonia in mainland China; on top of that his son was now hospitalized too. For young children to die from pneumonia is quite unusual, and this time the WHO inspectors decided to get samples of the virus themselves for analysis. At first their test suggested that the man and his son had indeed been infected with an avian flu, just as the Hong Kong officials had suspected, and it seemed that this might be a version of the 1997 strain from Hong Kong, but subsequent tests showed that the virus was unrelated to the 1997 strain.

With that news, Hong Kong health officials, as well as WHO experts in the region, became concerned that the avian virus in Hong Kong might be related to the cases of what was being called "atypical pneumonia" that they'd been hearing about in China. They were primed to look for such connections as part of their regular surveillance of possible new influenza strains. Then suddenly, on February 10, the Xinhua News Agency, mainland China's government-controlled news organization, published a strange announcement that an outbreak of atypical pneumonia was under control. The announcement also reported that plague and anthrax had been ruled out as causes, although it gave no explanation as to why these would have been suspected. The infectious agent causing the disease, the report said, was still unknown but was presumed to be some sort of virus, because the disease did not respond to antibiotics, which combat bacteria.

WHO officials pressed the Chinese government for more details about the outbreak. In response, the Chinese Ministry of Health informed the WHO that it had 305 cases, which included five deaths. It also mentioned, to WHO's surprise, that the cases had been mounting since the end of 2002. Becoming really worried now about a possible bird flu outbreak, disease specialists at the WHO and the U.S. Centers for Disease Control and Prevention (CDC) alerted U.S. government officials. On February 20, U.S. Secretary of Health and Human Services Tommy Thompson, Surgeon General Richard Carmona, and the new director of the National Institutes of Health, Elias Zerhouni, urged Chinese officials to keep the rest of the world informed about the progress of the disease.

The next day Liu Jianlun, a 64-year-old respiratory disease specialist at Sun Yatsen hospital in the Guangdong province, checked into the Metropole Hotel in Hong Kong to attend a wedding. He had been treating several sick patients and didn't feel well himself. Guests from Singapore and Toronto were staying on the same floor, as was Johnny Chen, a Chinese-American businessman based in Shanghai. Chen flew next to Hanoi, and by February 26 he was feeling extremely ill. He went to the Vietnam-France Hospital there, suffering from a high fever and a hacking cough. Doctors diagnosed him with pneumonia, but they had also heard rumors that a severe, untreatable respiratory infection was circulating among physicians in southeastern Asia, and they worried that Chen might have fallen ill with that. So the Hanoi doctors reported Chen's condition to the regional WHO office in Manila.

A veteran doctor there, Carlo Urbani, took special note of the alert, and rather than send a staff physician—the usual procedure—he decided to go himself.

Urbani, a balding 46-year-old from Italy, was an expert in human diseases caused by parasitic worms. But he was also a tireless foe of infectious diseases, and for years he had hunted down pathogens that were killing legions of people in the poorest corners of the world. For a time he was head of the Italian chapter of Doctors Without Borders, and in 1999, when that organization won the Nobel Peace Prize, he said that doctors had to "stay close to victims" if they wanted to wipe out disease. Urbani practiced what he preached.

When he arrived at the Hanoi hospital he was alarmed. Chen's fever was erratic, he wasn't eating, he was slipping into and out of consciousness, and he needed a ventilator to breathe. Urbani immediately took blood and saliva samples. He became even more anxious over the next few days when, one by one, doctors and nurses who had contact with Chen were falling ill with the same symptoms. Clearly, whatever Chen was carrying was a highly contagious infection. Urbani urged all the hospital's personnel to wear high-filter face masks and double-gowns, which were not routine in poor hospitals there or around the rest of the world, and told the hospital's administrators to isolate everyone who was ill. Then he sent an urgent alert to his colleagues at WHO headquarters in Geneva.

Urbani continued to monitor Chen, and doctors tried to ease his severe pneumonia, but Chen did not improve. By March 9 dozens of the hospital's workers had become gravely sick. Urbani and the director of the WHO Hanoi office met with Vietnamese officials and advised them to isolate patients and screen travelers arriving from Guangdong. On March 11 officials quarantined the entire hospital, and other hospitals in the area instituted strict infection-control procedures.

That same day, Urbani left on a plane for Bangkok, where he was scheduled to address a conference on the topic of deworming schoolchildren. But before he departed he called ahead to a fellow WHO doctor there to say that he was feeling ill himself. An American disease expert who was on assignment for the WHO in Bangkok went to the airport to pick him up. Urbani, feverish and ragged-looking, warned his colleague to keep his distance, and the colleague insisted Urbani get help, and brought him to a city hospital.

Urbani's condition was the last straw for WHO leaders, who had been monitoring developments surrounding the mystery illness for three weeks now. Part of the WHO mission is to act as the central clearinghouse for information about outbreaks around the globe, from Ebola and cholera to flu and TB. Member states, which include most nations, are supposed to forward breaking health information, especially about infectious diseases, to the WHO so that it can monitor all potential outbreaks. At this point the WHO realized that too many people in too many places in southeastern Asia were coming down with the same serious condition. The WHO alerted its Global Outbreak and Response Network, an international coalition of 100 national governments and scientific institutions that have expertise in infectious diseases. The mandate: Scour every bit of medical and media information about anything even remotely related to what was already known, and check samples from patients to figure out what was going on.

Then, on March 13, WHO officials got some truly troubling news. Doctors in Toronto reported that a 43-year-old man, Tse Chi Kwan, had just died of intractable pneumonia. The chilling detail was that his mother, Sui-Chu Kwan, who lived with him and his wife, had died at their home of the same illness one week earlier—right after flying back from the Metropole Hotel in Hong Kong. She had stayed in the same place as Johnny Chen and Liu Jianlun. The very same afternoon, Chen died of respiratory failure in Asia.

If that wasn't bad enough, nurses who had treated Kwan at a Toronto hospital were getting sick, repeating the pattern that had occurred in Hanoi. And German officials had just intercepted a plane during a stopover in Frankfurt to hustle off a 32-year-old doctor who was on board. The doctor had treated patients with similar symptoms in Singapore, and had then attended a medical conference at the Crowne Plaza Hotel in Midtown Manhattan. He, his wife, and his mother—all on the same plane—now had high fevers and rough coughs. WHO's global network was picking up all these reports and plotting a map of deadly infection on three continents. The mystery illness was quickly jetting around the world.

The WHO's analysis had been unusually quick. Widely criticized for its slow response to the AIDS epidemic a decade earlier, the WHO had made improvements. To a large extent the organization had always been, and still was, concerned primarily with chronic scourges like TB, malaria, and cholera, which require a well-planned, long-term response. The exception was influenza, which WHO monitored relentlessly. Was this new infectious disease in the same category? Top WHO officials, including David Heymann, the executive director of communicable diseases, called for an emergency teleconference of WHO officials and leaders from national health agencies such as the CDC. The group would have to decide if this new menace was truly a global threat—and if so, how to respond. Heymann would host the meeting from Geneva on Saturday morning, March 15.

Within hours, Julie Gerberding, director of the CDC, convened that agency's own infectious-disease experts at its brand-new Marcus Emergency Operations Center, a 7,000-square-foot bunker at CDC headquarters in Atlanta. The center had been quickly built after the anthrax attacks, when staffers found themselves communicating with nothing more than standard telephones and pagers. The center wasn't even officially opened yet, but it had been equipped with 85 workstations, nine team rooms, and a secure, state-of-the-art command station sporting high-frequency telecommunications networks and geographic disease-mapping systems. The technology assembled in this facility would enable Gerberding to communicate in real-time with the Department of Health and Human Services, federal intelligence and emergency response officials, and state and local public health officials, creating an ad hoc network among such groups that otherwise did not usually coordinate.

When WHO contacts from around the world met in Geneva on Saturday, March 15, they brought with them as much local intelligence as they could. Their reports revealed that the mystery illness was highly contagious—it was infecting so many people so quickly in Hong Kong, Singapore, Vietnam, and Toronto that it must be spreading by simple close contact—people coughing or sneezing on one another. And it had hopped continents so easily that people must remain contagious for some time, meaning travelers could spread the pathogen far and wide. WHO officials issued a statement right after the meeting ended saying that the illness was a "worldwide health threat"—a pronouncement WHO had never made at the outset of any new disease, not even AIDS. The representatives at the meeting also came up with a name for the mystery illness: "severe acute respiratory syndrome," which was quickly shortened to SARS.

The proclamation was more startling than controversial. Some health officials thought it was an overreaction, but the WHO was compelled by the pace at which SARS was traveling. History showed that wild outbreaks, like the flu pandemics that had ravaged the world in 1918, 1957, and 1968, could only be controlled with international coordination. And unlike in those eras, routine air travel could now fan the flames. WHO also now had the Internet and other reporting technology that allowed it quickly to chart the disease's spread, eliminating guesswork that in years past might have prompted the agency to take a wait-and-see approach. It was now possible to identify and respond to threats so much more quickly, and an early alert was justified.

The WHO's willingness to take decisive action based on incomplete information represented an important cultural change in the organization. Perhaps the person most responsible for moving the WHO into the fast lane was Jonathan Mann, who established WHO's Global Programme on AIDS in 1986. He had struggled to expand that program from an initial two-man operation to a staff of more than 200, but left in 1990 in a bitter policy dispute. He felt the WHO wasn't responding vigorously enough to AIDS, particularly in Africa, and was not doing enough to prevent further spread of the disease. Eventually the WHO came around to his point of view and began to act faster on news of outbreaks of all kinds. Unfortunately, Mann was killed in a Swissair crash over Nova Scotia in 1998—while on his way to Geneva for his first meeting at the WHO since he had left.

Less than three weeks had passed from the time Carlo Urbani suspected something was awry at the Hanoi hospital to the time the WHO declared a global emergency, an achievement of unprecedented speed in world disease detection. And yet, because of China's secrecy during the prior two months, SARS was able to get a foothold anyway across the world and was still spreading despite containment measures being put into place. Under intense international criticism for not telling the world sooner, Chinese officials finally began to offer up more information on what they were still calling a mystery illness. They didn't know why it suddenly appeared. It didn't respond to antibiotics or antiviral drugs. They believed the outbreak had started in the Guangdong province, and yet still insisted that it was tapering off and they had it under control.

The international infectious disease community was angered by China's lack of information, because they needed every bit of data they could get to figure out what was going on. It seemed that SARS might be the twenty-first century's first case of a danger that they were already deeply troubled about: that a sudden, genetic mutation in an infectious microbe would enable it to race through people around the world.

Fortunately, the global medical community had made remarkable progress in recent years in using genetic analysis to fight new microbial battles. By that point, however, labs in five countries had still failed to identify any known infectious agent as a cause of SARS. More tests were needed to determine if this "virus" was indeed a new form of a known pathogen, such as influenza, or a unique beast, and whether it had leapt from animals or not. Until these questions were answered, the medical community could only guess at the best way to control SARS.

The inability to pin down a culprit, suspicion about a possible cover-up in China, and heightened newspaper and television hype about a worldwide scourge stirred panic in some countries over the next ten days. Residents in Hong Kong rushed to buy protective masks. Health authorities in Hong Kong, Vietnam, Singapore, and China began visually screening airline passengers arriving from or departing from the countries where SARS had been reported as they got off airplanes; anyone who appeared to have SARS symptoms—fever or bad cough—was pulled aside, and many were forced into hospital isolation rooms. Though this was an unusually strong response that some deemed an invasion of privacy, these officials weren't willing to take any chances, and most of the individuals snared were frightened enough for their lives that they cooperated without complaint.

Meanwhile, a frenzy of international lab work was underway. The WHO had used its Global Alert and Response Network to distribute patient samples to twelve medical research institutions around the world, including the CDC, and they all began looking for telltale molecules that would indicate killer microbes.

Then, on March 18, scientists in both Hong Kong and Germany simultaneously reported that they had found what appeared to be virus particles in samples from three patients that seemed to belong to the paramyxovirus family, which is responsible for croup, measles, mumps, rubella, and several animal diseases. Shortly thereafter, the strain was identified as human metapneumovirus, a cause of pneumonia that was discovered in the Netherlands in 2001.[1]

But six days later, on March 24, Gerberding announced that CDC sleuths had found contradictory evidence that the germ was actually from the coronavirus family, named because of the viruses' characteristic crown-like structure. The samples they were examining included ones from Carlo Urbani himself. However, this strain appeared to be different from all known human coronaviruses.[2] A Hong Kong lab reached a similar conclusion the same day.

The medical world shuddered at the news—coronaviruses are the second most prevalent cause of the common cold. Virtually everyone has faced at least one bout of coronavirus infection by adulthood. These viruses usually infect cells of the upper respiratory tract but can infect nerve cells and immune system cells too. They go airborne when an infected person coughs or sneezes, and they can linger alive on surfaces from sponges and countertops to surgical gloves for several hours. They typically cause colds in adults rather than children, yet some microbiologists think the pathogens play a role in multiple sclerosis. If a new, more virulent form of coronavirus had evolved, the disease it was spreading would be highly contagious.

Scientists were also alarmed because a new type of coronavirus might well have evolved by leaping from an animal to a human. Various types of coronaviruses infect dogs, cats, rats, cows, pigs, turkeys, and chickens, and the human viruses are closely related. One study of sixty Japanese pig farms showed that 91 percent of the animals had been infected with the swine version of the virus, and 82 percent had been infected with what appeared to be a human coronavirus. While no evidence yet suggested that a pig and human coronavirus had recombined, that possibility had to be seriously considered. Indeed, the Chinese description of the town in which SARS was first noted indicated that people and pigs lived in close proximity there. When microbes make the leap from animal to human, they can be especially virulent.

The scientists had to consider another explanation: that a human coronavirus had simply mutated on its own to a more lethal form. There was a third possibility too; after consulting with Gerberding, Klaus Stohr, the virologist leading the WHO investigation, acknowledged that SARS could be caused by a co-infection of the corona and paramyxo viruses. After a week of analysis, the world medical community still could not definitively identify the causative agent. More people were falling ill, and increasing numbers were dying. By March 24, the WHO's global response network tally indicated that 456 people in 14 countries outside of China had come down with the disease, and 17 had died from it. Under continuing world pressure for information, China admitted that it had 792 cases, with 31 deaths so far.

The experts at the WHO knew they had to figure out why some patients developed more severe illness than others. Some people were recovering, though it usually took them several weeks. And while most of the deaths were in the elderly or those with underlying complications such as heart disease, young healthy people were dying too. Because China was the country of origin and had the highest number of cases as well as recoveries, the WHO requested permission from the Chinese government to send a team of investigators there, but China delayed access.

In the face of a rising menace, the WHO network convened a teleconference on March 26 to bring together 80 physicians from 13 countries so they could share their experiences in treating SARS patients. They pieced together a coherent set of symptoms and the first numbers about recovery and fatality rates. SARS usually began with modest to high fever, chills, muscle aches and a dry cough, they agreed. After about a week most patients began to get better, even if they received no specific therapy. But 10 to 20 percent found it increasingly difficult to breathe, and many required ventilators. A majority did improve after many days of struggle, but 3 to 9 percent of people who contracted SARS died, typically from heart or lung failure as breathing became impossible.

The incubation period seemed to be two to seven days, but could be as short as one day or as many as ten. Transmission seemed to be most likely by droplets coughed or sneezed into the air by contagious people, and it seemed that those particles could be contagious to people who were up to six feet away. A contaminated object, such as a doorknob or sink handle, might also transmit the germ if an infected person had grabbed it shortly before the next person came along. So far, no therapy had been effective for any patient whether they were mildly or fatally ill, and that was deeply troubling. Antibiotics were useless, and steroids and ribavirin—among the most prevalent countermeasures to viruses—also seemed to do nothing.

During that teleconference the role of the so-called superspreaders first came to light. In Hanoi, Johnny Chen gave SARS to more than half the health care workers who came into contact with him. The concept that certain people are more efficient at transmitting infections had been suggested in other epidemics—for example, men with genital infections that caused ulcers were thought to be an important trigger of AIDS in Africa. But the SARS superspreaders seemed to be far more infectious. No one had any idea what made them this way.

Carlo Urbani, the doctor who had first alerted the world to the disease, would have been intensely interested in what the doctors were saying during the teleconference. But he was fighting for his life in the Bangkok hospital. Two WHO specialists from Germany and Australia arrived to try to help him. They gave him ribavirin, but it did nothing. Growing patches of disease showed on his lung X rays. His wife, Giuliana, flew in from Italy to be with him, able only to talk to him across a double-paned glass window along one of the hastily rigged isolation room's walls. His lungs steadily filled with fluid. Despite being hooked up to a powerful ventilator, on March 29 Carlo Urbani died. His doctors said he had probably inhaled so much virus as he tried to treat so many people that he was doomed from the start.[3]

In the next few days the world learned about Urbani's fate. He had been the first to recognize SARS as a new and dangerous disease, and his strong warnings had convinced the WHO that it had to react swiftly and strongly. This champion of helping the most helpless had succeeded in alerting the world, saving countless lives. His quick action had limited the number of cases in the Vietnam-France Hospital and had prevented SARS from spreading across that country. The SARS virus was so infectious that even another week's delay in recognizing the problem would have spread the germ to hundreds if not thousands more people before control measures were put into place, one nation at a time. Carlo Urbani's funeral was held in Castelplanio, Italy. His wife and two of his three sons walked behind his casket.

By April 7, three weeks after the WHO announced a global threat, SARS had become a household acronym—and it had developed into a worldwide concern. The WHO and CDC were coordinating things closely as they updated the number of cases daily. Through its new emergency operations center, the CDC held a telebriefing nearly once a week for the media. With the WHO, it also produced unprecedented satellite broadcasts and Webcasts for clinicians and public health officials about clinical features, epidemiological investigations, and laboratory results. Information on an outbreak had never before been gathered and disseminated worldwide so efficiently.

The efforts to analyze the microbe that was causing SARS were also of unprecedented speed and scale. Researchers were becoming more convinced that the offending microbe was a coronavirus. The CDC and labs in Hong Kong, Singapore, Japan, Canada, Britain, the Netherlands, France, and Germany had reached this conclusion by searching for antibodies in blood from different patients that reacted with known pathogens, which would indicate a possible culprit. They had also searched for telltale bits of genetic DNA or RNA that might match gene sequences from known microbes; experts at a Japanese lab had ruled out all known influenza strains but had found no correlation to known pathogens.

Further comparison of a small bit of the microbe's genome was enough for the CDC to conclude that the pathogen seemed to be unlike any previously known coronavirus. This would be consistent with it causing a new disease. And yet the CDC was not ready to say that this coronavirus actually caused the disease, having learned from a previous experience in 1999, when it first identified West Nile virus as St. Louis encephalitis. In order to figure out whether this exact virus led to SARS, researchers in the Netherlands began injecting primates with it to see if it caused similar symptoms. Until they could show this, scientists could not devise a foolproof test doctors could use to diagnose patients.

Even if a mutated coronavirus could be definitively identified as the trigger for SARS, that wouldn't mean anyone had found a cure. Microbiologists were already working to sequence the pathogen's entire genome; a complete map would help them decipher what made the germ so deadly and so contagious and if it had a weak spot that drugs could target. The sequence might also give them a clue as to where it originated. However, the sequencing was likely to take several weeks, and developing a drug to attack the virus would take a good deal longer.

While researchers waited for the complete genetic analysis, they busily tried to devise tests that could at least indicate whether suspect pathogens were in a person's bloodstream. Such detection would allow health officials to identify and isolate people who were carriers earlier, slowing the illness's spread. By April 7 the network of labs had developed three rudimentary tests. Two relied on finding antibodies to SARS in a person's blood. The tests would help microbiologists figure out who became sick from the virus and who became a carrier without getting ill, as well as which people could infect others and for how long.[4] At this point, the medical community determined that the best method they had for detection was the rule of thumb that a negative antibody test in a potential patient early on, followed by a positive one about two or three weeks later, would be a sign of infection, and the person would be considered an official case. Likewise, a fourfold rise in the amount of antibody from one test to the next would qualify a person as a patient.

The third test used a technique known as the polymerase chain reaction to determine if a suspect coronavirus was present in a person's bloodstream. This technique duplicates a virus's nucleic acid and identifies it, a method that would be preferred because it could identify the virus within only five days after exposure. By contrast the two antibody tests did not seem effective until ten to twenty days after symptoms appeared.

These tests were the best scientists could do for the moment, and they didn't help the medical community much. Doctors would have to decide to isolate patients before such test results would be available, on the basis of symptoms like fever and cough, as well as whether a person had had contact with other SARS patients or had traveled from hot spots where the disease prevailed.

To provide whatever additional help to doctors they could, researchers also tried to figure out how the disease progressed in the human body, which would help with detecting it. They determined that the virus could be detected in secretions from the nose and mouth in the first five days of illness, and in the blood at about day 5. After that it moved into the lungs, causing dry cough, which would make detection more difficult, since a lung biopsy would be needed. They also thought the virus, or at least pieces of it, would appear in the feces from days 10 to 30.

As the research community continued to wait for the genetic analysis of the virus, scientists also began to think through the logistics of developing a vaccine or maybe even a cure. The first step in vaccine development would be to show that a sample of killed

SARS virus induced an immune response in lab animals and that it protected them from infection. Researchers were inspired by knowing that there were vaccines that protected animals from certain animal coronaviruses. Anthony Fauci, director of the National Institute of Allergy and Infectious Diseases, said his agency could perhaps develop a first-generation vaccine in about a year if it was lucky, but more effective vaccines would take longer.

Designing a useful drug was a long shot. The track record for drugs that could cure viral infections was not good. Fortunately, researchers had quickly found a way to grow the coronavirus in the lab. Using this method, the USAMRIID military lab in Bethesda, Maryland, began assessing a range of antiviral drugs that were on the market or in development; even though the coronavirus determination was not yet ironclad, it was the best anyone could go on.

On April 10, a WHO team that had finally gained access to Guangdong province in China on April 3 concluded its initial investigation, saying it was unable to pinpoint the origins of the disease. It did not uncover a link to birds, pigs, or other animals. No help there.

However, in a stunning feat of biotechnology, researchers at the British Columbia Cancer Agency announced just two days later, on April 12, that they had sequenced the genome of a coronavirus from one of the Toronto patients.[5] They had completed the genetic sequencing in six days of round-the-clock work, a phenomenal improvement over what would typically take several months. The next day the CDC announced it had sequenced a suspect coronavirus too, and on April 13 the agency posted the sequence on its website, along with an analysis of its relatedness to other coronaviruses. Reproduction and analysis of both genomes was to appear in the May 2 issue of *Science,* with a prepublication version available on the Web April 30, an unusual step.

High-level scientists analyzing the results concluded that the genome was different from all previously known human and animal coronaviruses. Marco Marra, director of the Genome Sciences Center at the Canadian cancer institute, said the culprit appeared to be the first in a new, fourth group of coronaviruses. Unfortunately, that meant disease hunters would have to search more widely to trace whether the microbe had jumped from an animal species, and if so, which one.

Finally, on April 16, the WHO got the scientific corroboration it needed to say that, yes, the new coronavirus caused SARS. A team led by Albert Osterhaus at the Erasmus Medical Center in Rotterdam announced that monkeys they had infected with the virus had developed a disease identical to SARS. Armed with this proof and the full genetic sequence, researchers could now search the virus for proteins that drugs could target as a vaccine or cure. That work would likely take much of 2003 and beyond. Marra did say his lab's map was a draft and that the various labs would have to cross-check their genomes to work out a final sequence. Sure enough, within days, comparisons showed that the sequences from the CDC and a Vancouver lab were nearly identical. Happily, this suggested the virus was not now mutating very rapidly. By April 29, seven additional full or partial sequences would become available on the Web, five of them from the Beijing Genomics Institute, but they would appear to be somewhat less related to the original two genomes that were sequenced. The microbe's origin would remain a mystery for the time being.

The response of the world medical research community to SARS was characterized by unprecedented speed and efficacy. The community demonstrated that it has evolved powerful procedures for combating such outbreaks. And yet, the still-unfolding SARS saga had exposed some troubling weak spots in world disease detection, control, and containment procedures. The fast-spreading epidemic largely incapacitated many hospitals, inflicted considerable damage on several national economies, and raised difficult questions about how much government control is appropriate and of what kinds.

Within two weeks after the first patient was identified in Toronto, health officials ordered two hospitals that had numerous SARS cases to limit their admittance of new patients to emergency victims and women about to give birth. Visitors were allowed only for dying patients, and everyone coming into the hospitals had to put on a mask. The surrounding Ontario province restricted the flow of patients and visitors in 200 hospitals. So many doctors and nurses were sick in certain hospitals in Hong Kong, Singapore, and China that the staff's ability to care for patients was compromised.

The range of public health responses was wide, and debate was fierce about which methods were helpful or ill-advised. Singapore, for example, quickly imposed extremely strong controls. Starting March 29, travelers arriving from Hong Kong or Beijing were greeted by medics wearing masks who asked questions about their health and people they had had contact with. Anyone thought to be suspect was taken away by ambulance for further evaluation. Hospital staffers were required to have their temperature taken twice a day. The country's schools were closed, and thousands of people were quarantined simply on the suspicion that they were carrying the disease.

Steps in Hong Kong were less stringent and slower in coming. Officials did ultimately quarantine an entire high-rise building in the densely populated Amoy Gardens complex after a number of its residents became ill. They worried that infectious particles were circulating through the building's air systems, or perhaps were being carried on elevator buttons, by roaches or rats, or in waste water leaking from the building's plumbing. Residents were moved to a makeshift camp. Then Hong Kong leaders closed all public schools indefinitely after a few children came down with the disease.

For the first time in its fifty-five-year history, the WHO recommended that travelers avoid certain parts of the world because of a disease. On April 2 the organization advised people to stay away from Guangdong province and Hong Kong unless travel was essential. Tourists and businesspeople dropped plans to travel to affected Asian countries. Intel cancelled two big conferences it was to hold in Asia. Companies such as UBS, the large Swiss bank, told employees who had been in Asia to stay home for ten days. Bookings at airlines from Continental to KLM Royal Dutch Airlines, already down due to fears of terrorism and the American invasion of Iraq, dropped even faster. The WHO then added Hanoi and Singapore to the no-go list. Malaysia became the first country to bar entry to Chinese nationals. Border guards required travelers from other hot spots like Canada and Vietnam to show health certificates saying they were free of SARS.

By April 21 there was no doubt that the economic damage inflicted on Hong Kong, Hanoi, and Singapore had been substantial. Visitor arrivals to Singapore had dropped 61 percent. The U.S. investment bank Goldman Sachs predicted Singapore's economic output would drop by half a percent in the current quarter because of SARS-related precautions, and Hong Kong's output would decrease by seven-tenths of a percent. Engineers who routinely flew between Taiwan and mainland China to keep up production of integrated circuits were being grounded; the region is the world's largest producer of computer microchips. Even so, the disease spread to Taiwan. Later, the Asian Development Bank said SARS could end up costing Asia $28 billion.

Economic consequences cropped up elsewhere. Restaurant and shop owners in Chinatowns in cities such as San Francisco and New York reported a 90 percent drop in revenues. Airlines cut more than 10 percent of all flights from the United States to Asia. Some economists even suggested that the SARS disruption could be the last straw that pushed the world into a global recession.

Then on April 23, the WHO issued an advisory that travelers should avoid Toronto for "all but essential travel," stunning the city's leaders, who had taken aggressive containment measures. They had already quarantined 7,000 suspected SARS patients or carriers. The damning declaration virtually closed down tourist and business travel to the city.

Some critics said the WHO's decision to issue the Toronto advisory was partly political, highlighting that the disease was a western problem too in order to help smooth things over with China. But the fact is that the rate of spread in Toronto was truly troubling. The city had 136 probable cases and the number was rising. Leaks in the system for tracking down and isolating people who might have had contact with carriers were also coming to light. An infected nurse had taken two crowded commuter trains home. Certain hospital workers or visitors were still coming down with the infection, despite careful hygiene and control procedures. As Donald Low, chief of microbiology at the city's Mount Sinai Hospital, told the *New York Times,* "This virus follows no rules or regulations of epidemiology. It's so highly contagious. There is no precedence for this kind of disease." The WHO felt it simply could not take any chances.

In addition to economic fallout, SARS forced nations to grapple with dicey social issues as they deployed new disease-detection devices and detention procedures. Airports in Singapore hastily installed infrared heat detection devices by the standard metal-detector stations; the inspectors aimed the instruments at passenger's foreheads as they passed by, and the detector showed in computer-enhanced color if the temperature of a person's face and breath was higher than normal, indicating they might have a fever. Hong Kong installed 215 infrared detectors at border crossings with mainland China. Toronto considered the scanners for its airports too.

By and large, passengers felt reassured by the detectors, though they didn't like the added delay in what was already a longer boarding process due to more thorough security screening since the September 11, 2001, terrorist attacks on the United States. Technology does not exist in a vacuum, however. As more infectious diseases threaten us, governments will have to decide whether new disease detection procedures are necessary, if they will cost too much, and if they impinge on civil rights. Cracking down on an epidemic with optimal speed may be impossible without invasions of personal freedom or privacy. Health authorities in this case have ordered people who are carrying SARS to stay home, and in the future they might require carriers to wear electronic ID bracelets so authorities can monitor their movements. When China finally decided to take SARS seriously, residents of Beijing were dismayed to find that the government had reinstituted the "snooping grannies"—groups of retired people who had spied on neighbors in the past to find out who was not living up to Communist party ideals. They were now watching to see if suspected SARS patients who had been confined to their homes tried to break quarantine.

In the United States, the uncomfortable questions posed by quarantine and the collection of detailed personal information from potential carriers has been largely avoided. Americans with SARS began walking into hospitals only after media and WHO alerts had been raised. But almost all of those patients had been exposed while traveling in other countries. Because they were quickly contained, and because the general state of infection control is much higher in American hospitals, the disease didn't spread fast or far. The CDC's Julie Gerberding also said the country was just plain lucky not to have had any superspreaders.

Nonetheless, the SARS calamity does highlight in a deeply thought-provoking way the need to establish clear protocols to govern control measures as we face future outbreaks. What would happen if superspreaders like Johnny Chen and Tse Chi Kwan turned up in packed Times Square in Manhattan on New Year's Eve, or at Disney World? Would Americans be willing to be quarantined if a new disease started in their country and the seriousness of the disease was still debatable? How much disruption of the economy would the government be willing to risk?

As frightening as SARS is, the virus is just the latest in an ongoing string of new pathogens that have evolved seemingly out of nowhere. SARS illustrates dramatically how quickly infectious diseases can arise and spread around the world. The outbreak teaches us many lessons.

First and foremost, we must realize that international travel is now so widespread and so fast that any disease can arrive anywhere at any time. The leading theory about West Nile virus is that a mosquito stowed away on an airplane and brought it to the United States. New diseases like West Nile virus and SARS can then gain a foothold before anyone realizes we are under attack. If the pathogen is highly contagious, it can spread in wider and wider circles before it has even aroused suspicion. The incubation period for most infectious diseases is days, and it takes only 36 to 72 hours to circumnavigate the globe, so most new diseases can arrive before anyone knows about them. As Marty Cetron, CDC's disease transmission expert, said at one of the SARS telebriefings, "The speed and volume of air travel poses a new challenge."

The second lesson is that it takes researchers a while to figure out how a new disease behaves. Revising initial thoughts about the contagiousness of the SARS virus, on May 4, a full six weeks into the outbreak, German scientists announced they had discovered that the microbe can survive on plastic surfaces—such as elevator buttons—for up to 24 hours. They also found it can live in human feces for as long as four days, complicating toilet sanitation in hospitals and homes. Such information is vital in knowing what guidelines to issue to world health organizations, but it still takes considerable time to gather.

It can take even longer to determine what exactly a new disease is. The exercise is not easy. First, astute clinicians have to realize that they are looking at something different. For SARS that person was Carlo Urbani, but several more days might have gone by before someone with his acute awareness picked up on the disease. The next step in identifying an emergent disease is that lab clinicians have to examine patient samples to find the actual organism, or antibodies to the organism. Often, as happens with global diseases such as tuberculosis and hepatitis C, the microbe hides inside the human body, and a test that shows signs of unusual immune system activity, like a TB test, is needed for a diagnosis. Even then, fine tuning might be needed. It took specialists three years to come up with the right culprit for AIDS, seven years for hepatitis C.

Thanks to the crush of international lab work in the case of SARS, scientists came up with primary suspects and preliminary tests for the disease in only a few weeks. Yet several months would be needed to make the tests reliable enough for use on the general public. Even then, it would still be possible for a person to be negative to all three of the preliminary tests, but still have SARS if the virus had moved deep into the lungs. This conundrum is what makes it so difficult for doctors to determine exactly what variety of pathogen among many is creating a pneumonia. Doctors usually start antibiotics immediately just in case it is bacterial, yet the cause of half of all pneumonia cases is never identified.

On a larger scale, the SARS microbe provides disturbing evidence that even when microbiologists have sequenced a genome, that alone may not tell them whether a pathogen jumped from animals, or if a previously unknown human strain simply mutated or picked up genes from other organisms. This information is vital in knowing how best to develop a drug to combat a disease.

Did SARS arise because two viruses joined up to cause more damage than either one alone? That's what WHO lead investigator Klaus Stohr feared. And that's why the role of a possible paramyxo virus had to be determined. Both the paramyxo and corona pathogens are respiratory viruses. Perhaps the superspreaders like Johnny Chen were carrying both viruses. Medical journals describe historic incidents of superinfectors such as Typhoid Mary, who transmitted typhoid fever slowly but widely in New York City after 1906, or a medical student who spread drug-resistant *Staph aureus* bacteria to dozens of hospital patients in 1996. A leading theory is that these people are unusually contagious because they have a second infection that somehow powers up the first. Fear of the consequences of commingling viruses grew tremendously in mid-April when SARS arrived in South Africa, where AIDS runs rampant. Would HIV co-infection prompt an even more devastating expansion?

Thankfully, the response to the SARS epidemic has been swift and sure. One reason is that the WHO had set up the Global Outbreak Alert and Response Network. The network got its start in April 2000 in Geneva when more than 100 nations and nongovernmental health institutions like the Red Cross and Doctors Without Borders agreed to work together to combat the spread of natural and bioterrorist outbreaks. Part of the plan is to provide countries with technical assistance in identifying the causes of unusual infectious diseases, their sources, and their routes of transmission. As soon as the WHO declared SARS a worldwide health threat it quickly coordinated the distribution of patient samples to twelve laboratories that would try to identify the cause. Then it swiftly convened the March 26 teleconference so doctors from all continents could share their experiences. Julie Gerberding, the CDC's director, said the cooperation had been "extraordinary." She also said the fact that a diagnostic test could be derived so quickly was "an unbelievable achievement in science" which she credited to the close collaboration among the world's labs.

The CDC also sent more than a dozen investigators overseas who, with WHO staff, would look into the disease's origin and spread. At the same time, it handed out hundreds of thousands of health alert notices to travelers returning by plane to the U.S. directly or indirectly from mainland China, Hong Kong, Vietnam, Toronto, and Singapore. CDC agents directed people who seemed to be ill to hospitals. The CDC recommended infection control procedures to hospitals—steps that would limit SARS from spreading from patients to nurses, doctors, and family members. Meantime, President George Bush added SARS to the list of diseases for which a person could be quarantined, giving the public health community another important tool with which to control the disease.

It takes strong political will to impose a quarantine that is aggressive enough to really curb a disease's spread. Although Hong Kong quarantined people from one building at the Amoy Gardens apartments, it didn't close off other buildings in the same complex, even though patients were found in those buildings as well. Ironically, the concern for civil liberties can become a health liability when officials delay implementing public control measures until the evidence is so overwhelming that possible violation of civil rights can be justified. Making the decision about when to impose quarantine is difficult, and officials must have guidelines in place so that they'll be able to cope with such emergencies as soon as they occur.

The trickiest issue is how a country's culture affects public willingness to support tough measures. The leaders and citizens of places like Singapore, Vietnam, Hong Kong, and Beijing have proven willing to impose and accept quarantine, and even martial law, due to their civic histories. The same may not be true for countries in other parts of the world.

When the anthrax letters surfaced in October 2001, U.S. quarantine laws were suddenly exposed as being incredibly antiquated. Some had been unchanged since they were originally written 200 years earlier, when whole towns were wiped out by smallpox and other diseases. In response to an open request by the Bush administration, James Hodge at the Center for Law and the Public's

Health at Johns Hopkins University, and Lawrence Gostin at the Center for Law and the Public's Health at Georgetown University, drafted model legislation that would give a state's governor and its health administrators broad powers to impose quarantines in the event of a large-scale medical emergency. That law would also grant the president the power to issue quarantine orders if a disease was expected to cross state borders. Since then twenty-two states and the District of Columbia have passed parts or versions of the legislation, and other such laws have been drafted. But there has been little public debate on this issue, and SARS fears may hasten the passage of quarantine laws in the remaining states before the public has weighed in.

Existing laws might also be broadened to apply to any perceived medical emergency. But opponents of this new quarantine law such as George Annas, chairman of health law at Boston University, note that many public health officials are political appointees who have little training in health issues, and worry that they could be swayed into ordering mass quarantines or vaccinations with unproven medicines. They also fear that officials could abuse the powers. When the AIDS epidemic was growing fast in the 1980s, calls arose from the general public to quarantine gay men because they might spread AIDS. Those cries actually ended up being counterproductive because they led to a distrust of the health establishment and resistance to testing for HIV among gay men for years.

Most of the new laws being proposed say that people under quarantine must receive food and proper medical care, and that their confinement must be by the "least restrictive means." Timely hearings would be required for people who challenged orders or conditions of their confinement. And yet most of the laws allow officials to put people in quarantine without their seeing a judge or a doctor. And even if officials force people to take vaccines they don't want, and they develop health problems or the vaccine kills them (this is a risk with some vaccines), they or their families cannot sue anyone to recoup losses or for damages.

A separate issue is the effectiveness of quarantines. Not all people in Canada or Hong Kong obeyed voluntary quarantine, a condition that can prompt authorities to detain people in isolated locations where they can be watched more easily.

Despite imperfections, the coordinated action by the international health community once the word got out was impressive. Professional and national rivalries were set aside to protect one and all. This model of international cooperation, preparedness, and commitment of resources to global health should be applauded. And it should be institutionalized, repeated each time another serious infectious disease breaks out.

Just as important, if the CDC and its counterparts abroad are to be able to react effectively, legislators in various nations must protect these agencies from budget cuts, even during financially difficult times. The same goes for state and local health departments, which must turn the CDC's recommendations into action plans at each local hospital, clinic, and town. The new killer diseases are unpredictable. No one knows when the next influenza pandemic will strike, when another hemorrhagic fever like Ebola will pop up, or when another SARS will erupt.

For its part, the CDC seems to have learned a lesson about informing the public. Unlike its close-mouthed investigation of the anthrax attacks, this time it informed the media and hence the public almost daily. Panic did not ensue. The CDC was also highly reactive in informing doctors; it quickly posted information on its website telling physicians how to recognize SARS, what precautions to take in treating suspect people, and the grave importance of rapidly informing the public health sector of potential cases, including the CDC.

What did not happen, in the beginning, was good international communication. China's initial concealment of SARS was shameful. Governments the world over cannot quietly wait to see if emerging outbreaks will "burn out." That's what the British government was criticized for doing with the human form of mad cow disease. On April 4 a top disease prevention official in China did apologize for "poor coordination" in keeping the world and the Chinese public informed about the rise of SARS. He promised the government would make public more information and would create a national disease-warning system.

And yet, even with that rare admission, the lesson seemed lost. On April 9 a senior but retired physician from the Chinese military bravely spoke out, saying the national health ministry was lying about how many people were hospitalized with SARS in Beijing. The health ministry had said there were twelve cases and three deaths, but the doctor said that the number of patients in military hospitals alone was on the order of 100. Another nonmilitary doctor then went public, indicating his hospital in Beijing had dozens of cases. In China, leaders tend to keep health statistics secret under the guise of preventing panic, but observers there say officials are particularly tight-lipped about trouble in Beijing because it is the nation's capital. At the time, it was hard to know how much of this was willful, but some China observers suggested that the primary motivation for the cover-up was that a new premier had just assumed power and the government didn't want a crisis to interfere with that transition. The only other explanation was that the medical system was simply poor at finding and reporting unusual illnesses—which was no more reassuring.

Political pressure did finally work with China. On April 17 an international WHO team formed by the organization's Beijing office said local authorities had significantly understated the epidemic's severity. Hearing this, government leaders from around the world angrily called upon China's ruling party to intervene. On April 18 Communist Party leaders suddenly declared a nationwide war on SARS and ordered officials to stop hiding the extent of the scourge. President Hu Jintao warned that the disease could affect the country's development and stability. With this impetus, Hu's government disclosed on April 20 that cases nationwide, and in Beijing in particular, were many times higher than had been reported—a rare admission of government failure. Party leaders also let it be known that they had replaced national health minister Zhang Wenkang and Beijing mayor and deputy party chief Meng Xuenong, implying that it was bureaucratic failure rather than Party policy that had caused the reporting problems.

China's silence hurt the rest of the world, but it hurt China worst of all. In a live, televised news conference on May 1 in Beijing, the new acting mayor, Wang Qishan, said the SARS epidemic there was severe and "uncontrolled." The city's caseload had leaped by 101 in a single day, to 1,440. Wang said the city had been woefully unprepared, had misled the public, and now faced a shortage of hospitals and staff trained to handle SARS cases. As a result, not all suspected patients could be hospitalized in a timely fashion, a hint that perhaps it was indeed dangerous for the public to mingle unprotected. When Wang had taken over on April 20 he instituted stringent control measures, and by May 1 some 9,000 city residents had been put in isolation. Students and migrant workers were barred from leaving the city. But the damage had been done; tens of thousands had already fled Beijing, spreading the virus across the country, notably to rural areas where public health resources would be weak or nonexistent.

Countries are sometimes reluctant to report outbreaks because they don't want to harm tourism and international business dealings. But such reluctance is simply unacceptable. So many people travel so quickly throughout the world now that there is no longer such a thing as a "local outbreak" of an infectious disease. The rapid spread of SARS from Guangdong province and then Hong Kong is the ultimate proof. As the *New York Times* wrote in a March 18 editorial, "This is not the way to be a good neighbor in an age when germs are only a jet flight away." According to the CDC, each year more than 80 million nonresident passengers arrive in China and 13 million arrive in Hong Kong. Approximately 460,000 residents of China, Hong Kong, and Vietnam travel to the United States. As Marci Layton, assistant health commissioner in New York City, told the *Times,* "It's a global village, and we're at its door."

The world community must work to combat the problem of governments that are not forthcoming with disease information. During the Senate hearing, Heymann was asked if something could be done in the future so nations did not delay in providing information about outbreaks. Heymann pointed out that the WHO has no power to override any country's sovereignty. It can only pressure a nation's leaders.

With luck, that might change. The World Health Assembly was scheduled to meet May 19. On the agenda was updating the International Sanitary Regulations, which were enacted in 1951 and require only the reporting of cholera, plague, and yellow fever to the WHO. The assembly's goal was to transform that code into a new global treaty by 2005. Delegates had been revising the regulations for eight years, but Heymann believed SARS might finally give the group the momentum needed to pass the treaty this year. If passed, it would require signatory nations to maintain at least a minimal surveillance system, and to report to the WHO any "public outbreak of international concern" whether or not an infectious agent has been identified.[6]

At the same time, Heymann admitted that the WHO communicable disease branch was "very underequipped to respond" to such notices and was scrambling to reassign personnel to meet the SARS challenge. He said the branch only had 29 people and had to depend on help from other agencies like the CDC, which had more than 300 staff assigned to SARS.

Even if top government and health officials are not secretive, they can fail to alert the world due to poor judgment if they are overly sanguine about the foe they face. When the WHO did declare SARS a global threat, many high-ranking observers still thought the disease might just flare out on its own. But as the different experiences in Vietnam and Hong Kong showed, it took vigorous public health measures to stop the epidemic. Health officials had to track down people who were ill and find other people who had close contact with them and then quarantine those people. They would also have to watch national borders so infected people would not walk into or out of their country. Even Canada failed to screen outgoing passengers until the WHO imposed the travel advisory against Toronto.

It can indeed be difficult for government and health officials to determine the appropriate degree of fear about a new disease. Threatening diseases make big stories, and the media whips up the frenzy day after day. That is why national health organizations must take a forceful public relations lead in maintaining the right balance between preaching caution and spreading worry. Part of that goal, too, is to make sure that the calamity of the moment does not steal attention away from other infectious diseases that statistically are greater threats to our lives. More people die in a week of influenza in the United States than had died worldwide of SARS during the outbreak's first two months.

Nonetheless, in the SARS case the attention was warranted, because if the virus continues to spread around the globe and become endemic, many more people will die. SARS may not cause more deaths than flu, but it could be more disruptive in terms of closing down hospitals and disabling health care delivery. And the disease is continuing to evolve and could become more deadly.

That danger was brought into sharp relief on May 2, when Denis Lo, a chemical pathologist at the Chinese University of Hong Kong, announced that SARS had already mutated into a second form. Researchers there who had compared genetic sequences of virus samples from eleven patients had found that two forms had arisen in Hong Kong alone. "We have shown that the SARS coronavirus is undergoing rapid evolution in our population," Lo told the Associated Press. It was unclear whether the second form caused different patient symptoms, however. WHO scientists tempered the news by noting that the coronavirus family is prone to mutations, so the finding was not a surprise.

Yet even if it did not change the disease's contagiousness or severity, the mutation could complicate development of a diagnostic test, which is based on specific genomic sequences. And if the mutation affected the shape of proteins on the virus's surface, that could hamper creation of vaccines, which train a person's immune system to recognize a particular set of protein shapes. Researchers would also have to determine if people whose immune systems naturally beat one SARS strain would have developed immunity to the second; if not, they could get SARS from the mutant strain all over again.

Because SARS was not immediately contained, it spread too far ever to be eradicated without incredible long-term effort, if at all. And because it mutates, it is almost certainly here to stay, adding to a growing panoply of infectious threats. The only real debate is whether SARS will become a relentless foe, like influenza, or an ailment that breaks out sporadically, like Ebola does thus far. In either case, if the world does not take measures to eradicate any new disease, it will surely become endemic. Work toward vaccines and cures must continue unabated. SARS could become more of a nightmare because it is so highly contagious. All you have to do is breathe it in. To get AIDS, by contrast, you must be exposed to an infected person's bodily fluids, and even then infection is not likely. Every ride you take in an airplane, commuter train, subway car, or elevator could expose you to SARS if it became widely disseminated.

Half-hearted measures about new diseases can be perilous. A lack of vigilance caused New York City to dismantle its tuberculosis clinics in the 1980s, only to see TB surge back stronger than ever, and in drug-resistant form to boot. It's also what many people said about West Nile virus in 2000 and 2001, the years after the pathogen attacked New York City. Aggressive spraying of pesticides and manual clearing of standing water where mosquitoes breed did keep the disease down, but West Nile came roaring back, with more than ten times the cases and deaths nationwide in 2002. It certainly had not burned out. The American public, and people in general, tend to become alarmed, possibly even overly so, when a new disease first rears its ugly head, but too often they forget all about the threat as soon as deaths seem to subside. As the CDC's Julie Gerberding told the Senate, the rise of SARS was just "the beginning of a problem." Infectious diseases are evolving fast. SARS is just the latest. There will be more.

During the CDC's April 4 teleconference, James Hughes, director of the National Center for Infectious Diseases, put the implication of SARS in blunt terms. "This is a fire drill," he said, adding that, "If we have a bioterrorism attack, we will be doing many of the same things that we're doing now, operating through multidisciplinary, headquarters-based, and field teams."

In the rush to combat SARS, that larger perspective may have escaped the public, but not the country's top health experts. Three days after the world learned of SARS, the U.S. Institute of Medicine released a comprehensive report prepared by a wide-ranging group of eighteen medical and scientific experts about the threat that infectious diseases poses to the nation's people.[7] On the very first page the institute declares that "the impact of infectious diseases on the United States has only increased. . . . The present reality is that public health and medical communities are inadequately prepared to recognize and respond to microbial threats, particularly those originating elsewhere."

Although the institute crafted the report with no knowledge of SARS, its pages are eerie as they describe how new pathogens can emerge, as if the experts had already watched the SARS story unfold. The report describes thirteen factors that account for new or reemerging threats, and SARS exemplifies them remarkably. Lead factors include the "tremendous evolutionary potential" of microbes, changing climate, altered ecosystems, increased human contact with animals, new medical technologies that have created novel pathways for the spread of infections, and "the rapid and virtually unrestricted transport of humans, animals, foods, and other goods, which can lead to the broad dissemination of pathogens and their vectors throughout the world." In the face of these factors, the institute concludes, all the world's people "ultimately share the same global microbial landscape."

Indeed, the institute experts write, "Any of these factors alone can trigger problems, but their convergence creates especially high-risk environments where infectious diseases may readily emerge, or re-emerge. . . . It is conceivable, in fact, that in certain places microbial 'perfect storms' could occur, and unlike meteorological 'perfect storms,' the events would not be on the order of once in a century, but frequent."

The institute's recommendations for facing down the new killer diseases also closely echo the conclusions presented by experts in this book.

As of the time of this writing, SARS had already exacted a terrible cost in human lives. By May 17, the WHO had reported 7,761 cases, including 623 deaths, in 28 countries on 5 continents.[8] Of those, 66 patients (though no fatalities) were in the United States; 5,209 cases and 282 deaths were in mainland China, and another 1,710 cases and 243 deaths were in Hong Kong. Although the WHO had noted that the epidemic might show signs of subsiding, it remained vigilant. Closer examination of cases indicated the overall death rate was 15 percent—much higher than first expected—and was more than 50 percent in people over age 65 who had been hospitalized.

The next time a newly evolved disease breaks out in the world, the cost could be even higher. During the short weeks of the SARS outbreak, other pathogens were breaking out across the world, yet they were lost in the news. An Ebola outbreak in Africa went into its fourth and fifth months. A new strain of avian flu was wiping out chickens across Belgium and the Netherlands, had already caused 79 human cases of conjunctivitis, as well as several cases of a flu-like illness that was responsible for pneumonia that killed a 57-year-old veterinary surgeon. More than 19 million birds had been slaughtered. And world health experts feared that if this avian flu recombined with a human flu virus, a pandemic could erupt that would dwarf the current SARS epidemic.

CHAPTER ONE

THE CASE OF JEANNIE BROWN

As the case of SARS demonstrates, the human race is in the midst of an escalating biological war against an army of microscopic foes that we have seriously underestimated. These infectious microbes are awe-inspiring in number and are stunningly resilient. They can reproduce in hours or even minutes, mutating and adapting far faster than we humans can. They are learning to outmaneuver our immune systems and our drugs, and they are rapidly evolving deadly new capabilities. There is really no way to keep these pathogens out of our homes or workplaces. The threat they pose is as formidable as any that the human race has ever faced, and despite all our brilliant advances in science, pharmaceuticals, and medical technology, these simple organisms could still get the better of us. We have no idea when the next Ebola virus or mad cow disease might emerge in our backyard, or when some ordinary microbe might swap genes with a deadly germ to create a superpathogen even more virulent and quickly spreading than AIDS. On any given day one of these mutant diseases might strike down any one of us, or someone we love, so we must teach ourselves how to be on high alert. For these diseases can be terribly deceptive. Just as SARS mimics a cold or the flu in its early stages, other new killers can fool patients and doctors into overlooking their danger.

Jeannie Brown was mad at her live-in boyfriend. She wouldn't confront him because that would cause a scene.

Instead she figured she'd get him back just a little. When she made him a sandwich for his lunch one day, she left the plastic wrapper on the slice of cheese. Then she wrote a little note saying "Ha ha!" on his napkin, and crumpled the brown bag closed.

This small rebellion was about all Jeannie cared to muster that steamy June morning. She was weary. Until a few weeks ago she'd been working two full-time jobs: the night shift at the Handy Pantry convenience store in Canton, North Carolina, her hometown, and then from 7:30 A.M. to 4:30 P.M. pulling patient files at Blue Ridge Bone and Joint in nearby Asheville, which did hip replacements and orthopedics.[1] Sometimes she became so exhausted she'd fall asleep in her car right in the Blue Ridge parking lot. She was also depressed about having gained weight, and she was struggling with her self-esteem. On top of that, her doctor had just started her on medication to reduce high blood pressure, and she was only thirty-two! She had also had some kind of laryngitis for almost all of April.

What Jeannie wanted most was to take off for some rest at Myrtle Beach. She knew her best friend, Vivian, was heading there for a three-day weekend with her family and friends. Jeannie hadn't been there in three years, and the young summer was already hot. Though her complexion was naturally dark, she loved it when her round face was deeply tan. So she decided she'd surprise Vivian and just show up Friday morning. Her boyfriend would be away for several days anyway, and she knew her mother, Audrey Trantham, who lived nearby, would look after her ten-year-old son, Cameron. Just the thought of the getaway raised her spirits.

Jeannie left her yellow, double-wide mobile home in the fog-shrouded Smoky Mountain foothills before dawn the next morning and slipped across the interstate in her old, white Nissan Pathfinder. She reached the beach by 10:30 A.M. The blue sky was shining brightly. She crept along the beachfront streets, looking for Vivian's car. Just as she found it, Vivian appeared from a small, cedar rental house. She spotted Jeannie right away and came running over, delighted to see that Jeannie had done something so spontaneous. For once Jeannie was taking control, doing something for herself.

After Jeannie went inside and changed into her bathing suit, she reappeared, rubbing her neck behind her left ear.

"Are you okay?" Vivian asked.

"Yeah, I guess," Jeannie answered. "It feels like I have a knot in my neck. I must have slept wrong."

Vivian, tall like her friend, looked at Jeannie's neck, but didn't see any redness. When she touched the muscle she didn't feel any tightness, either. They figured the soreness would work itself out, and, looking forward to a relaxing day on the beach, they pulled back their ash-blond hair and stepped out into the warm sun.

The two women sat on the busy beach most of the day, reminiscing about the good times they'd had together. Jeannie talked about Cameron, and about her stressful relationship with her boyfriend. She also talked about how much she loved the ocean, and about how some people preferred to be cremated when they died, so their ashes could be spread out over the water.

The pair agreed that they and Vivian's family should go out to dinner that night. Jeannie suggested they go to the Outback Steakhouse, and Vivian was fine with that idea. When they got there, though, the line at the boisterous restaurant was long, and Jeannie worried that Vivian's relatives would be exasperated at the wait. But everyone was in a relaxed mood, and didn't care. The group had a good time, but Jeannie turned quiet during dinner and mostly sat staring into space. Vivian thought she might be hurting and just wasn't telling anyone. Sure enough, when they got back to the house at around 10:00 P.M., Jeannie went straight to bed.

On Saturday morning Vivian emerged from her bedroom just as Jeannie was stepping out from her room across the hall. Jeannie was holding her shoulder, obviously in pain.

"The knot's moved down to my shoulder," she said with a wince.

"Is it sore?" Vivian asked.

"To tell you the truth, I hurt all over."

Vivian still saw no discoloration or lump. But when she touched Jeannie's shoulder, Jeannie yelled, "That hurt!"

Vivian gave her some Aleve, and Jeannie said she was going to lie back down in bed. Though Vivian told Jeannie she was worried,

Jeannie shooed her off to the beach. But when Vivian returned at lunchtime she found a note from Jeannie saying, "Thanks for everything. When it comes to pain, I guess I'm a wimp. I'm going back home to rest in my own bed."

During the five-hour drive back to Canton, Jeannie's pain grew worse. When she arrived home she didn't feel like calling her mother to have her bring Cameron home. She needed some rest, so she just closed the drapes and fell asleep on the couch.

The next morning, Jeannie's sister-in-law, Sherry Trantham, got a surprise visit from Jeannie. Sherry, a slim thirty-seven-year-old, worked as a nurse at the Silver Bluff Village nursing home, three miles from Jeannie's house, in the knotty hills across from the local paper mill. Normally Jeannie wouldn't impose, but now she was worried something was seriously wrong with her, and she hoped Sherry could do something.

Sherry's eyes opened wide behind her round glasses when she saw Jeannie's left shoulder. It was swollen, reddish purple all over, and warm to the touch.

"What happened?" she asked, alarmed. Her eyes narrowed. "Did somebody hit you?"

"No," Jeannie said. "I don't know what's wrong."

Sherry sensed that Jeannie was afraid. Usually smiley and talkative, she was quiet, and clearly had been crying. Sherry's first thought was that Jeannie's collarbone was broken. She knew Jeannie hated hospitals and didn't like taking medication, and that because Jeannie was financially strapped, she fretted about having to pay bills her insurance wouldn't cover. But Sherry insisted that Jeannie go to the local clinic right away.

When the doctor at the clinic examined Jeannie's shoulder, he told her she had probably pulled a muscle, and gave her some medication for the pain. But by Sunday evening, still alone in her house, Jeannie began vomiting. She took Compazine to try to control the nausea, but it didn't work. Her shoulder was also now going into spasms. She was racked with fits of vomiting all that muggy summer night, and by Monday morning she was panicked. Surely one of the orthopedists she worked for at Blue Ridge could help. She called the office and said she would be there by noon.

Noon came and went. At 1:00 P.M., Cassandra Ledbetter, another file clerk, happened to look out the front window and saw Jeannie's Pathfinder pulling into the parking lot. Jeannie stumbled out into the pouring rain. She took slow, deliberate steps toward the building's slate entryway, focusing all her strength to stay upright. When she reached the big glass doors she just stood there, motionless and drenched. Cassandra and two other staff women rushed to help her inside. In the fluorescent light, Jeannie's office-mates were startled. Jeannie's face was pasty, and her stringy hair was matted to her head. Her eyes were distant; she gazed like a zombie across the waiting room, as if she were stoned. The twenty-minute ride from her house, a backroads shortcut, had taken an hour and a half. Five times she had pulled over to vomit out the door of her car. She reeked. Tears streamed down her pale face.

The women hustled Jeannie off to a doctor, who told Jeannie he thought she must be badly sunburned, and gave her medication to reduce the inflammation. Then he walked her out to the lobby to find someone to take her home. One of the nurses offered, but Jeannie refused, saying she would take the backroads again. She would be okay. Cassandra thought she knew what Jeannie was thinking, though she disagreed with Jeannie's decision. When you're a woman working full-time, trying to hold together a family, taking care of aging parents, trying to make ends meet, and dealing with all the usual discomforts a woman has to cope with, you just put up with the pain and try to keep going. Two hours later one of the nurses went to Jeannie's house to check on her, but Jeannie wouldn't let her in.

Jeannie's boyfriend was still gone, so at about seven-thirty that night, Jeannie managed to call Vivian. She blurted out her story between loud sobs. "I went to the clinic yesterday. . . . They said I pulled a muscle. They gave me Vicodin for the pain, but it's not working. . . . I saw the doctors at work this morning. They thought I had a sunburn. They gave me steroids, but they're not working, either. . . . The pain's so bad I can't stand it. . . . I am so red, Vivian."

Vivian was baffled. She and Jeannie tanned all the time, and she knew Jeannie hadn't been burned that day at the beach.

"Maybe you're having an allergic reaction to all the medication," Vivian said.

"I don't know."

"Jeannie, can I help you somehow?" Vivian asked with concern.

"No. I guess not," Jeannie said. "Thank you for being such a good friend." Then she hung up.

That night Jeannie once again threw up repeatedly, but she didn't go to the hospital. On Tuesday morning at 9:45 she called Sherry at the nursing home in desperation.

"Sherry, I'm dying," she cried. She was panting, extremely short of breath, and could barely speak. "Can you come down here?"

Sherry opened Jeannie's front door at 10:35, and the smell from inside the hot house stunned her. She found Jeannie lying on her bed, in clothes crusted with dried vomit. Her mouth and ears were crimson, her chest nearly black, her stomach, back, and thighs mottled purple. She looked as if she had been beaten nearly to death, except for the bottoms of her feet, which were snow-white. Her head was hot, but her toes and fingers were cold. Her nail beds were violet. Sherry's heart raced at the sight.

Jeannie's eyes, which everyone always said were brilliant crystal blue, were dull and glassy, but Sherry could see the struggle behind them. Tears ran down Jeannie's cheeks. She seemed cogent, but she was hyperventilating, and she pleaded for water. Sherry quickly took Jeannie's temperature. Leaning over Jeannie's face to see the thermometer, she was surprised to discover it read only 101. Then Jeannie heaved loudly. Thick, rust-brown blood spewed from her mouth, all over her bed and all over Sherry.

Sherry ran to the bathroom to get some towels. That's when she saw it; there was brown, vomited liquid in the hallway, on the bathroom walls. Sherry raced back to Jeannie and found she was now having a diarrhea attack, heavy with blackened blood.

"We've got to get you to a hospital," Sherry insisted. "I'm calling the rescue squad."

"No, don't," Jeannie gasped. "Sherry, you're a nurse, you can take care of me here."

"There's no way. I'm calling the rescue squad."

"No. I don't want to make a scene in the neighborhood."

"Who cares! You need to see a doctor," Sherry demanded.

"It'll just mean more bills."

"I'll pay for the ambulance!"

Sherry couldn't believe that Jeannie was still resisting. Was she delirious? Unsure what to try next, Sherry called Jeannie's mother, who was at work across town.

"Audrey, this is Sherry. We've got to do something for Jeannie. She says she's burning up, but her hands and feet are ice-cold. She keeps wanting water; she can't get enough of it."

"Call 911," Audrey said.

"She doesn't want me to."

"Call it anyway!" Jeannie's mother demanded.

Jeannie begged Sherry for more water while Sherry got her dressed. When the rescue squad arrived they immediately put Jeannie on a cardiac monitor. Incredibly, her heart was racing at 170 beats a minute, yet her blood pressure was so low they couldn't get a reading. They hurried Jeannie onto the gurney and took off down Highway 110 for Haywood Medical Center, twelve minutes away.

The ambulance raced up the wooded hill to the brick hospital, wind-driven rain pelting the windshield. It halted at 11:30 A.M. at the automatic double doors of Emergency.

Audrey had driven straight from her job to Haywood, and was waiting at the ER. As the EMTs lowered the gurney from the ambulance, Audrey could see the distress in her daughter's eyes. The EMTs had put Jeannie on oxygen, but she was fading out. As they rushed Jeannie past her, Audrey said, "Jeannie, I'm here, and I love you."

Four ER nurses quickly gathered around Jeannie and whisked her into a room. Audrey stood outside the door, clutching her brown knit sweater, her tall frame rigid in concentration. She heard Jeannie moan, then vomit. Then she overheard the nurses saying, "Come back, Jeannie, stay with us, stay with us. Your mom's right here."

The ER team paged Michael Keogh, the attending physician on call. Keogh was stunned when he saw Jeannie. Her face and body were a mottled mix of black splotches, which he figured were from local bleeding or damaged tissue. Yet her skin was also marked by tinted yellow patches, which to him indicated a lack of blood to the surface of the skin. She was losing consciousness, and he had to keep waking her to ask her questions. He thought she had been badly beaten, but she was lucid enough to deny it.

Realizing Jeannie was undergoing serious trauma, Keogh barked out orders as he wheeled an EKG machine to her side. As the EMTs had discovered, her heart rate was very fast, but strangely her blood pressure was also dangerously low. It was as if her heart was beating frantically to try to pump more blood, but didn't have the strength to push it. Keogh realized that Jeannie's circulatory system was failing; she was shutting down. She needed fluids immediately. It was too late even for an IV line. Keogh thrust a catheter into Jeannie's clavicle, below her neck, to pump fluids directly into her primary veins, and then injected dopamine into the fluid to get her blood pressure up. Now he had to figure out what was happening to her. He drew blood from her and sent it to the lab for emergency tests.

Keogh next went to talk to Audrey and Sherry, who were in the nearby waiting room. He told them he didn't know yet what was going on, but Jeannie's heart was not beating strongly enough to send oxygen through her body. "I have to sedate her," he told them, "and insert a tube to help her breathe." He asked Audrey to sign a release form for the emergency procedure, and asked Sherry about the medications Jeannie had been taking. When he heard she had been taking Vicodin, he thought maybe she had overdosed on it, and he asked Sherry to drive to Jeannie's house and bring back all the medicine bottles she could find. Keogh then returned to Jeannie, added a sedative to the fluids she was getting, thrust an air tube into her trachea, and fed a nasogastric tube down her throat to pump her stomach. Having done all he could to stabilize her, he told the nurses to get Jeannie to the intensive-care unit right away.

The bloodwork report reached Keogh about an hour later, at 1:00 P.M., the same time Sherry returned with the medicine bottles. The report showed safe levels of all the medications, which agreed with the evidence Sherry had brought in. But the report also showed that Jeannie's blood was turning to acid; the pH level had sunk from the typical 7.40 to 7.05, usually a sign of toxins from a rampant infection. Keogh was perplexed, though, because her white-blood-cell count—the standard indicator of infection—was relatively normal. The ratio of newly forming white cells, 52 percent, was a bit high but not extraordinary. If Jeannie had a heavy infection, her total white cell count would be much greater. Then Keogh noticed a smoking gun: Jeannie's level of creatinine was three times higher than normal. That meant her renal system—her kidneys—wasn't functioning. Something bizarre was going on inside this woman's body. Keogh suspected a massive infection at this point, but he had no idea what it was. He did know Jeannie needed a higher level of care than he could provide. He would have to send her to Mission St. Joe's, the big regional hospital in Asheville.

Matters got immediately worse, however. Suddenly Jeannie lapsed into cardiac arrest. Keogh worked furiously to stabilize her heart, which he managed to do. Then he ordered the ER staff to add another antibiotic to her fluids in case she did have some kind of infection, and to prep her for the ambulance. The drive to Mission St. Joe's would take a dangerous forty minutes—a life-or-death trip—but there was no other way; Mission's helicopter was stranded in the mountains, held down by the windy thunderstorm that was now soaking the entire region.

Keogh left Jeannie to place an emergency call to Dr. Kathryn Freyfogle, a twenty-year kidney specialist who saw patients at Mission.

"Are you on call?" he asked.

"Yes."

"I have a critically ill young woman who is in shock. She's in circulatory collapse, and she has abnormal renal failure. She's not even urinating. She's already been ventilated. Would you accept her in transfer?"

"Yes," Freyfogle responded.

Keogh just hoped he could get Jeannie there in time. He found Jeannie's mother in the hall outside Intensive Care to explain the plan. "We're going to move Jeannie to Mission. She must have some strange illness. We don't know what it is. Go on ahead to Mission and get her signed in."

The EMTs worked frantically to keep Jeannie alive during the tense drive, the catheter pumping salt water and dopamine into her veins, the tracheal tube blowing oxygen into her lungs as all of her systems shut down. By the time they arrived, Jeannie's heart had stopped. Even as they rushed her through the ER doors into Intensive Care, they performed the Advanced Cardiac Life Support Protocol (ACLS), the standard emergency procedure to try to revive a person who has gone into cardiac arrest. Dr. Freyfogle was waiting there with green scrubs on.

Freyfogle was prepared for the worst, but even so, when she saw Jeannie she was taken aback. Jeannie's entire body was now a mosaic of black, purple, and yellow. Her torso was one continuous bruise. The skin on her upper back and neck was blistering, seeping brown pus onto the gurney sheets. The whites of her eyes were bloodred, and she appeared comatose. She did not respond to anything the nurses were doing. Certainly she was in profound shock.

"Continue ACLS!" Freyfogle ordered. Four ICU nurses and two respiratory therapists surrounded Jeannie while Freyfogle slapped an EKG monitor on her. The instrument showed a bizarre pattern; Jeannie's heart raced along, then stopped cold. Then raced again, and stopped again. She had no pulse, which meant that despite her heart's furious efforts, blood was not circulating through her body. Freyfogle's adrenaline surged. She quickly added Lidocaine to the catheter mix, to try to stabilize Jeannie's heart rhythm, and steroids to try to restart her kidneys. Over the next half hour the crew managed to get a faint pulse a few times, but it would last no more than about thirty seconds and then disappear.

After forty-five minutes of frenzy, Freyfogle lowered her head. She knew that Jeannie was never going to pump blood or breathe again, and even if by some miracle she did, by now she would be brain-dead. At 5:50 P.M. on Tuesday, June 20, 2000, Dr. Kathryn Freyfogle called out, "Stop."

Freyfogle had not yet met Audrey or Sherry. She sent a nurse to bring them to a private office. When the nurse asked them to follow her, Audrey looked at Sherry in fright. Everything was moving so fast. She didn't know what to expect. She thought perhaps they were going to tell her that Jeannie had to have open-heart surgery, or that she had an aneurysm. The nurse left Audrey and Sherry standing in the small room, and Freyfogle walked in. She somberly looked Audrey in the eye and said quietly, "Mrs. Trantham, I'm so sorry. I just had to pronounce your daughter dead."

"Oh my God!" Audrey exclaimed. She stood at attention, staring straight ahead into space, in shock. Freyfogle paused, then swallowed hard, knowing that she had to say one more terrible thing. "We don't really know why Jeannie died," she said slowly. "Could there have been foul play?"

Audrey didn't even hear the words, unable to process what Freyfogle had just told her. Sherry looked at Freyfogle and shook her head.

"Mrs. Trantham," the doctor said, "I'd like to do an autopsy. But I need your permission."

Audrey didn't move or blink. She just said, "Okay."

Talk was hushed among the few family members and friends who gathered early at the cemetery that Thursday. Audrey waited at home as late as possible, because she didn't want to be at that graveyard one minute longer than necessary. Then the phone rang. It was Dr. Freyfogle. "Hello, Mrs. Trantham," she began. "I'm sorry to intrude. But I know how anxious everyone is to hear why Jeannie died. And as a doctor, I also think everyone who saw her in the last few days had better know . . ."

Audrey's husband drove her through the overcast morning to the burial ground. Audrey got out of the car and the others circled around her. Sherry thought Audrey seemed strangely distracted and asked if something had happened.

"Dr. Freyfogle called," Audrey said. "She had the autopsy results." Audrey winced in disbelief as she uttered her next words. "Jeannie was killed by an overwhelming infection. It destroyed her kidneys and her heart. The doctor said that by Tuesday it was unstoppable. She said she thought it was just Jeannie's day to go to heaven."

Once the disbelief cleared, anxiety set in. *I was with her for a whole day,* Vivian thought. *We used the same lipstick. We shared the same toothbrush! Did she give it to me?* As soon as Vivian got home she called her doctor. She was worried about herself, her husband, her friends, and her mother. They all had been with Jeannie on the beach.

Audrey was worried about herself, but even more about her ill, elderly mother, and about Jeannie's young son, Cameron. Sherry sweated beneath the dark blouse she had worn to the funeral. As a nurse, she knew better than the others just how easily infectious disease can spread, and she had taken the brunt of it the morning of Jeannie's death. *Jeannie threw up blood all over me,* she worried to herself. *She had diarrhea all over me. I must have it.* She feared for her seven-year-old daughter, Amber, who loved to hug her close every night. *Did I give it to her, too?* And what about the nursing home patients? They were easy prey for disease.

The autopsy had been a revolting experience for Freyfogle. It showed that Jeannie was killed by a wild infection doctors call "invasive Group A streptococcus."[2] The media call it "flesh-eating bacteria." The ravenous bacteria had turned Jeannie's kidneys into a mass of jellylike pus, her heart into a squishy sponge. A microbial war had been raging inside her body for at least a week, the rapidly multiplying bacteria knifing through muscle, organ, and blood-vessel cells while a few meager white cells tried to tackle them. The pain of tissue being consumed must have been excruciating. Jeannie's left shoulder muscles, where she had felt a "knot," were eaten relics of flesh. Toxins had spread throughout her body, the by-products of the rampaging bacterial feast. Her dying kidneys could not filter the poisons from her blood, and her constant vomiting was her body's vain attempt to expel them. As the toxins rose, Jeannie's blood pressure sank, and capillaries already being decimated by the bacteria gave way, releasing more blood beneath the skin, adding to the black and purple mottling and turning her eyes red.

By the morning that Sherry got Jeannie to Dr. Keogh, Jeannie's arteries and veins had become so pulped by the acidic poison that they leaked like perforated hoses, causing her blood pressure to plummet. Her circulatory system was so limp it couldn't get blood to her feet, and they turned white. It couldn't bring enough oxygen to her tissue or her brain, so she was forced to hyperventilate to try to get more. Freyfogle figured that the cascade of failures threw Jeannie's cardiovascular system into confusion, her brain into delirium. Near the end, the bacteria had begun to eat the skin on Jeannie's upper back and neck, causing the vile strep to ooze out into the open. And yet, Freyfogle discovered during further analysis of the Haywood blood cultures, the infection might have succumbed to the antibiotics penicillin or clindamycin if massive doses had been administered to Jeannie the week before. But no one would have suspected, then, that the vicious strep was present.

What is truly frightening about Jeannie's awful death is that invasive Group A strep is a subtle variation of the same germ that causes everyday strep throat in millions of people each year. In most individuals, the immune system can beat back the pathogen. But in others, for some reason, it cannot, and the invaders reproduce wildly, consuming muscles and organs, killing the victim by toxic shock or multiple organ failure—or both, as in Jeannie's case. Group A strep came to national attention in 1990 when it suddenly killed Jim Henson, beloved creator of the Muppets. One-third of the people who contract flesh-eating bacteria die, and the number of fatal U.S. cases has tripled to 2,000 annually in the last eight years. Why this common bacteria, which we all breathe in from other infected people, is turning increasingly deadly is a mystery.

The flesh-eating variant is only one of a growing gang of invasive Group A strep organisms that attack 10,000 to 15,000 people in the United States each year. In most cases they invade blood, muscles, or lungs but are contained. Medical researchers don't know why they run wild inside some people and not others. They appear to strike at random. You could inhale one of these germs on any given day. If your immune system can't effectively fight it, and if you don't see a doctor immediately who is keen enough to diagnose it and start you on antibiotics right away, it could ravage your body, just the way it did Jeannie Brown's.

In the past, Group A strep has fueled a number of deadly epidemics, including the scourges of rheumatic fever that swept the United States in the early 1900s, killing thousands of children a year. But by 1985 the number of rheumatic fever cases had declined to a few dozen annually. The medical community assumed that antibiotics were holding the rheumatic strep down. That complacency was shaken in the late 1980s when rheumatic fever suddenly rose among children and young adults in cities across the United States.[3] Then came reports of a raucous strain of strep that was sending adults into toxic shock, followed by the invasive flesh-eating bacteria. After nearly a century of steady remission, Group A strep had reemerged in horrible new killer forms.

The rise of deadly strep is just one small part of an alarming escalation in the evolution of infectious diseases. Scores of other bizarre organisms are springing to life. Since the 1970s—less than a blink in the millions of years that microbes have been evolving—experts have identified twenty old pathogens such as TB and cholera that have reemerged stronger or spread wider than ever, and thirty newly discovered diseases, including Lyme disease, West Nile virus, Ebola, and AIDS, and now SARS.[4]

Lyme disease was first described in 1977.[5] Today deer ticks are spreading the bacteria steadily, infecting more than 16,000 people a year. Originally confined to the Connecticut region, the disease has now cropped up as far away as northwest California. Its first symptom may be a characteristic bull's-eye rash, but when left untreated it can develop into chronic, debilitating arthritis. The same ticks can also transmit Powassan virus, which causes dangerous, brain-swelling encephalitis. Meanwhile, the backyard mosquito is spreading West Nile virus. After its surprise arrival in New York City in 1999, West Nile affected a few dozen people a year, but in 2002 it expanded at an alarming rate, infecting 3,698 people in forty-four states by mid-November, killing 212.[6]

Other killers, previously limited to far-off jungles, increasingly threaten us as they learn how to thrive in the temperate climes that blanket America, Europe, and Asia. The vicious Ebola virus from Africa weakens a person's blood vessels, causing a gruesome death by bleeding from the inside out.[7] Another tropical demon that causes massive internal bleeding is the Marburg virus. In 1967, Marburg escaped Africa and infected thirty-seven people in Germany and Belgrade, killing a quarter of them. The source was found to be monkeys that had been imported from Uganda in order to make polio vaccine. Though no large-scale outbreak of Marburg happened in that case, odds are, sooner or later the exponential increase in international travel will almost surely bring such diseases to our doorstep. North America was almost tested in February 2001 when a Congolese woman, traveling through Newark Airport in New Jersey and on to Canada, suddenly became extremely ill. Doctors feared she had brought Ebola or Marburg with her. Luckily that case turned out to be a false alarm.

These exotic diseases highlight the threat of species-jumping, whereby pathogens that have previously not infected humans figure out a way to make the leap. Once they do so, they may take decades to determine how best to feed off the human body and resist immune attacks. In the interim they may lie low, undetected, and then break out. This is just what happened with AIDS, which jumped from chimpanzees to humans. Beginning as early as the 1950s, the HIV virus made several quiet sorties to test the human species, slowly teaching itself how to more efficiently spread and beat our defenses. By the 1980s it had triggered a global pandemic. AIDS has now killed more than 15 million people worldwide (3 million in 2002 alone) and infected another 40 million, and has begun raging in such countries as China and India.

Sexually transmitted diseases such as gonorrhea are rising again after two decades of gradual decline, and dangerous ones, such as chlamydia, are zooming. Other classic scourges are strengthening, too. Malaria kills more than a million people a year. It is caused by a microscopic parasite named *Plasmodium* that travels back and forth between the liver and the red blood cells. Such parasites are another huge class of pathogens that can cause as much trouble as some bacteria and viruses. As *Plasmodium* bursts open large numbers of blood cells, it induces fevers and chills. Ultimately it can lodge red cells in the brain's blood vessels, blocking circulation and causing death. Mosquitoes inject the bugs into our bloodstream when they bite us. There are a few malaria drugs, but *Plasmodium* has largely figured out how to resist them. Malaria is not just a threat in poor countries; it attacks 1,000 to 2,000 people in the United States each year.

Infected food presents great problems, too.[8] There are now more than 76 million incidents of food poisoning a year in the United States alone. The bacteria *E. coli* O157 is just one of the rogues, causing everything from mild diarrhea to fatal poisoning. A whopping 80 percent of chickens now harbor campylobacter, a bacteria more prevalent than *E. coli* or salmonella, which infects 2.4 million Americans annually and kills more than 500. In thousands of others it leads to Guillain-Barré syndrome, which often results in temporary paralysis. The food-borne bacteria listeria causes fewer infections but even more deaths, as well as miscarriages. In October 2002, Pilgrim's Pride Corp. recalled 27.4 million pounds of its Wampler brand of ready-to-eat chicken and turkey products for possible listeria contamination—the largest food recall in U.S. history. Listeria can trigger high fever, severe headache, neck stiffness, and nausea, and can be fatal in young children, the elderly, and the frail.

Old demons that the Western world thought it had eradicated, such as plague, are back as well. And "stealth viruses," including herpes and hepatitis, are silently slipping into our cells and multiplying, then lying dormant, tricking our immune systems into overlooking them—only to burst forth periodically, causing fatal illnesses such as cancer and perhaps even heart disease.

Tens of thousands of Americans die annually from infectious diseases. Millions die worldwide. Unfortunately, we have developed vaccines for only a handful of them, and more and more of our drugs are becoming outmatched. Old organisms like staphylococcus and the bacteria that cause tuberculosis which we thought we had under control are rapidly evolving to resist our drugs—so effectively that microbiologists now talk about the possibility of a post-antibiotic era when those overprescribed drugs no longer work at all. And entirely new forms of pathogens, like the prions behind mad cow disease, are infiltrating and decimating our brains.

If all that weren't enough, the October 2001 anthrax attacks in the United States instantly turned what had long been viewed as the possibility of bioterrorism into a reality. The scare revealed how vulnerable we are to a terrorist group that gets its hands on anthrax or, even worse, smallpox, which is highly contagious and kills one-third of its victims. The Centers for Disease Control and Prevention (CDC) and Central Intelligence Agency (CIA) keep their eyes on a half-dozen other pathogens that can kill with abandon. Recently released reports also indicate that even more deadly organisms could be genetically engineered in high-tech biolabs.

Menacing as bioterrorism is, the danger from naturally evolving diseases is far greater. While we watch out our front door for terrorist germs, killer microbes such as the SARS virus are sneaking through our back door. The sheer number of pathogens is rising, and their diversity, reproductive vigor, and ability to mutate are daunting.

How could a young American woman die today the way Jeannie Brown did? You'd think that if a thirty-two-year-old with a bacterial infection got to a hospital, with all its drugs and high-tech machines, doctors could save her. Why didn't the doctors at the clinic, Blue Ridge, Haywood, or Mission know what was really wrong? A pulled muscle. Sunburn. An overdose. The physicians were far off. Why were they confounded? Maybe more attention or humility early on would have caused them at least to suspect they had a strange case on their hands, and prompted them to consult with experts. But the plain fact is that the flesh-eating form of strep is sufficiently uncommon and has roared onto the scene so fast that many doctors don't know how to identify it. Or they simply aren't on the alert for it. Many aren't trained in how to recognize new and reemerging infectious diseases, and others haven't kept up well enough with new developments after leaving training. By the time they diagnose an aggressive organism, the chances of saving a victim may be slim.

One problem is that many of the new killers can be deceptive. The early symptoms they cause are similar to those of more common ailments. Jeannie's fever and muscle aches would more typically be due to flu, and her vomiting to food poisoning. Low blood pressure can be caused by various heart problems. In Jeannie's case, her normal white-blood-cell count stumped Dr. Keogh; it should have been much higher if her immune system was trying to battle an infectious attack. Perhaps because the strep had attached so fast, the white cells were only beginning to ramp up. Or her bone marrow, which produces white cells, was shutting down as her body faded. By the morning of the day she died, even massive doses of antibiotics would probably have been too little, too late.

Not only is the medical community unable to recognize some of these diseases early enough, but the basic science has a long way to go. Researchers are at a loss to explain the mechanisms by which many of the deadly pathogens operate, and why they are more deadly for some people than others. No one knows what makes one strain of strep cause a sore throat and makes another dig deep into our flesh, bringing extraordinary destruction and toxic shock. Did the strep bacteria that killed Jeannie disguise itself so her immune system would not recognize it as infectious? Perhaps Jeannie was missing an enzyme or gene that would have helped her immune system kill the intruder; none of the people in contact with Jeannie became ill. Did the several strep throat infections she had had as a child leave her more vulnerable? Maybe a fault developed in her immune system after she had chicken pox, which she didn't get until she was an adult and was pregnant with Cameron. An investigation in 2000 by Boston Public Health Commission researchers of a 1997 invasive Group A strep outbreak in a Boston child-care center suggested that recent infection with chicken pox makes an individual more susceptible to Group A strep.[9] Or maybe the persistent cold in Jeannie's throat in April marked the beginning of a strep infection that for some reason she couldn't clear. The Boston researchers also found evidence from a 1992 outbreak in a Swedish child-care center that pharyngitis, a strep infection of the upper throat membrane, makes individuals more susceptible to invasive Group A strep and hastens the disease's spread.

Perhaps Jeannie's immune system was extremely weak after months of exhaustion, lack of sleep, and emotional stress.[10] Did the mix of medications in her last days suppress it even more? Or did this infectious disease just overtake Jeannie so quickly that her immune system never had a chance to muster a response?

Why didn't it kick in faster? Doctors often simply can't answer these questions.

Even more unsettling is that, to this day, no one can say how Jeannie became infected. Did Jeannie get it from one of the patients at Blue Ridge Bone and Joint, who may have been exposed to it while in the hospital for surgery? Jeannie could have picked it up through one of the paper cuts she got regularly from pulling files at work. Did she get it at the Outback restaurant, or the beach?

Maybe Cameron brought the virulent strain home from school, where kids carry strep regularly. Maybe some infected guy who came into the convenience store late one night coughed all over his money just before handing it to her at the register, and she rubbed her eyes after he left, trying to keep herself awake.

Jeannie Brown's case also highlights the unsettling reality of how little public-health officials would be able to do to stop the spread of a strep A outbreak. Indeed, Jeannie's autopsy results triggered a cold sweat among infectious-disease specialists at North Carolina's Public Health Department and the federal Centers for Disease Control and Prevention in Atlanta. Six months earlier, on December 30, 1999, Paula Clement, a twenty-nine-year-old woman from Swannanoa, North Carolina, only twenty miles from Jeannie's hometown of Canton, had died the same rapid death, also only hours after entering a hospital. Then, in February, thirty-one-year-old Terri McClelland of Black Mountain, only five miles east of Swannanoa, had been diagnosed with flesh-eating bacteria, too. Luckily for her, she had read a newspaper article about Paula Clement, recognized early that she was developing similar symptoms, and got herself to a doctor who immediately put her on six different antibiotics. After her body waged a two-week battle, she survived, but she required months of physical therapy to regain her muscle strength.

Had the three young mothers crossed paths? Had they acquired the deadly bacteria from some common source? After a newspaper article in the *Asheville Citizen-Times* recounted Jeannie's death and that of Paula Clement, a scare swept the area. It was then midsummer, when many residents make the trek east to the North

Carolina beaches. Rumors were flying that that was where Jeannie had picked up the disease. And she must have brought it back to Canton and Asheville. Which local stores did she shop in? Which bar might she have had a drink in? Audrey's mail lady asked Audrey's next-door neighbor if Jeannie's house had been fumigated. Scores of people called their doctors: Could they get it, too? How would they know if they had it? Friends called Vivian and Sherry, saying, "You were with her near the end. Aren't you afraid you're going to die?"

Examiners at the CDC in Atlanta ultimately determined that the cases of Jeannie Brown, Paula Clement, and Terri McClelland were unrelated. Thankfully, no outbreak seemed to be in the works.

From the early 1900s through the 1970s, the numbers of deaths from infections had been dropping decade by decade. Smallpox had been eradicated. Antibiotics had been developed that could beat down bacteria within days. We thought we had infectious diseases on the run. Big cities like New York dismantled expensive tuberculosis-control programs because TB seemed to be waning. In 1978 the United Nations predicted that by 2000, in "even the poorest nations . . . infectious diseases no longer would pose a major danger to human health."[11]

Then things started to go wrong. The AIDS epidemic took off. TB reemerged in a more deadly form that resisted multiple drugs. Infections had dropped to an all-time low in 1980, but by 1998 the annual U.S. death rate from infectious diseases doubled to 170,000. Globally that year, infections ranked first on the list of the world's killers, taking out 15 million people.

Bad luck? Hardly. The onslaught is happening largely because we've brought it upon ourselves. Germs are mutating faster than ever because we are helping—or forcing—them to change. We push economic development into the jungle, rousting exotic pathogens. We pump greenhouse gases into the atmosphere, altering the planet's weather patterns, which facilitates the spread of disease-causing organisms. In June 2002 scientists reported the first evidence that global warming was accelerating outbreaks of new and more virulent diseases in a range of plant and animal species, and that the same warming could trigger a similar acceleration of diseases in humans.[12]

More of us are traveling to foreign countries, where we may pick up exotic microbes and carry them back to our families and friends. This is why SARS spread so rapidly. We import more food from countries with poor safety standards. More parents choose not to immunize their children against "old" diseases, which has allowed measles, mumps, and whooping cough to rise to more than 6,000 U.S. cases annually.[13] We overprescribe antibiotics and overuse antimicrobial soaps, killing only the less dangerous pathogens, which eliminates the natural competition that prevents the more threatening ones from growing even stronger. And yet we focus drug research on compounds that might cure the big-ticket diseases such as cancer and heart disease, while doing too little for infectious diseases; in 1999, the U.S. Food and Drug Administration approved thc first new type of antibiotic in thirty-four years. We cut funding for public-health measures against a given disease as soon as it slips below epidemic levels, only to allow it to reemerge, better able to flout our immune systems and resist our drugs.

Mostly we have succumbed to our own hubris. We have underestimated the microbes. The examples are chilling. Scientists have been startled by the recent discovery that the friendly *E. coli* bacteria that aids our digestion has picked up killer genes from other nasty bacteria and begun to poison us. It is as if a house cat assimilated a snake's gene for making venom and turned on us as we slept. The variant *E. coli* O157 now commonly contaminates food. If you eat it, the strain can cause severe bloody diarrhea, kidney failure, and death. Outbreaks have flared up everywhere, from a water park in Atlanta to a coffeehouse in Seattle.

Defiant germs are persisting in even the cleanest environments. For several months in late 2002, Norwalk virus broke out repeatedly on cruise ships run by Carnival, Holland America, and Disney despite intensive scrub-downs between voyages. More than 1,400 passengers experienced vomiting and diarrhea on several oceanliners.

"Common" ailments have also become really dangerous.

Despite the availability of flu vaccinations, an average of 36,000 Americans still die from the influenza virus every year,[14] and microbiologists fear that the flu is overdue to change abruptly, sparking another pandemic like the 1918 "Spanish flu" that killed 500,000 people in the United States alone.

The bizarre pathogens known as prions—normal proteins that become destructive after they strangely misfold—are rising, too. In Britain, people who have eaten prion-contaminated beef have been dying, years after cows themselves began falling from mad cow disease. The first U.S. victim, twenty-three-year-old Charlene in Florida (her last name has not been released), was diagnosed in October 2002 after having lived in Britain until she was thirteen. And now it appears that a similarly insidious pathogen may be spreading in the United States. "Chronic wasting disease" (CWD) is slaying a growing number of deer and elk across the central United States, boring mad-cow-like holes into their brains. The animals are eaten by hunters and others. Could this disease also spread to humans? In the summer of 2002, scientists feared they had discovered the first human deaths from CWD: a hunter in Wisconsin and his friend from Minnesota who had shared a wild game feast, and another hunter in Montana.[15]

We must also be alert about a number of stealthy pathogens that lie dormant inside our bodies, only to erupt years later, which are spreading vigorously. The HIV virus that causes AIDS mutates more rapidly than the flu, can develop drug resistance against all the licensed AIDS drugs, and has learned how to destroy the immune system that initially fights it. We still have not discovered an effective vaccine. Tuberculosis, herpes, and hepatitis C each have evolved to avoid destruction by drugs or the immune system. Some pathogens have even figured out how to package drug-resistance genes and trade them, to speed up their collective ability to outmatch us.

The good news is that a veritable army of professionals is working feverishly to fight the pathogen threat. Thousands of microbiologists and geneticists within academia and biotech companies are trying to understand better how deadly microorganisms work—and how to defend against them. They are making important discoveries that could lead to more targeted drugs or even genetic cures. Some experts have spent their entire careers unraveling the secrets of a single organism, as Rod Welch at the University of Wisconsin has for deadly *E. coli*. Welch's work shows how the development of molecular and genetic research over the last ten years has revolutionized our ability to find ways to defeat our microscopic enemies. Today graduate students can routinely clone genes and begin to figure out how they work. Microbiologists have decoded the entire genome of *E. coli* O157, HIV, and the parasite that causes malaria, among scores of others. Now they can begin to apply this knowledge to design imaginative treatments.

As we saw in the case of SARS, the CDC is a leading force in supplying intelligence, preventive techniques, and cures for the pathogen war. This impressive organization of 8,500 scientific, medical, and public-health experts operates around the country, directed by a core staff from a twenty-eight-acre campus in suburban Atlanta. One of the CDC leaders fighting in the trenches is Nancy Cox. She is combining genetic work with good old gumshoe investigating to support her nerve-racking job of monitoring unusual flu strains that arise regularly around the world, so she can advise the World Health Organization (WHO) on which strains to include in a given year's flu vaccine. Pharmaceutical companies then have to gear up quickly to get the vaccine to market in time. The lives of thousands depend on Cox getting it right.

Since the anthrax attacks, Congress has given the CDC much more money to battle bioterrorism, and the work will spill over to improve our defenses against all infectious diseases, as it did in the response to SARS. Part of that is a nationwide network for reporting patients who are under attack from terrorist-delivered or naturally occurring pathogens, and for advising doctors in every locale what warning signs to watch for and how to respond. The CDC is expanding a variety of Internet networks to help.

Physicians, meanwhile, are working better with local and state health departments to improve early detection of emerging outbreaks of unusual diseases. For example, the alertness and coordination demonstrated by Deborah Asnis, a physician in the Queens borough of New York City; Tracey McNamara, a veterinarian at the

Bronx Zoo; and Marci Layton, the city's assistant health commissioner, led to the recognition that West Nile virus had reached our shores. Their concerted response probably saved many lives. The Working Group on Civilian Biodefense, a collection of physicians and epidemiologists, has been publishing a series of articles in medical journals to teach doctors nationwide how to recognize unusual infections—like tularemia, one of the most infectious pathogenic bacteria known—which are high on the list of potential terrorist agents, and what the medical consensus is for the most appropriate treatments.

The army of defenders also include researchers who double as medical activists, like Stuart Levy, who is waging a campaign of public information to improve our health system. Levy, a professor of molecular biology at Tufts University, runs the Alliance for the Prudent Use of Antibiotics, which tries to explain to citizens and doctors how our overuse of antibiotics is creating a network of superbacteria that we may never be able to defeat.

In all of this, an educated public is a crucial component—whether it's knowing when to take two aspirins and rest, when to see a doctor fast, how to protect oneself against mosquitoes, or understanding the importance of vaccinating children. We can empower ourselves greatly by becoming better informed about the threat. Providing readers with a detailed look at the range of diseases we must be aware of and what experts are doing to combat them is the mission of this book.

CHAPTER TWO

THE BIOTERRORISM CHALLENGE: *Meeting the Threat*

Though naturally evolving pathogens are the focus of this book, evaluating what we learned from the 2001 anthrax attacks provides a sobering introduction to the challenges we face in fighting the pathogens teeming around. The anthrax scare was a wake-up call, revealing how vulnerable the United States is to terrorists who might decide to spread a killer organism. But the episode also spoke volumes about how unprepared we are to fend off the rising tide of natural microbe outbreaks.

In the aftermath of the scare, Congress allocated $2 billion to national disease-control agencies to improve the country's ability to detect and react to biological threats. The extra funds paid dividends immediately, helping to pin down the surprisingly rapid spread of West Nile virus in 2002. But that level of cash infusion was a one-shot deal. And the uncomfortable irony is that continuing to reallocate massive resources to counter bioterrorism could actually jeopardize our already underfunded defenses against naturally evolving threats. Determining the optimal balance of resources will be one of the public-health community's great challenges in the coming years.

Stepping up our detection and warning systems and our development of vaccines and cures is a complex undertaking. We are still not ready to handle an anthrax offensive, much less a smallpox attack, which would be far more destructive. And bioterrorists can exploit many other organisms that are just as deadly. The techniques also now exist to create entirely new pathogens. In July

2002, scientists at the State University of New York at Stony Brook announced to the world that they had built a synthetic polio virus from scratch.[1] They ordered stretches of DNA through the mail from a commercial supplier and stitched them together following a model of the real pathogen's genetic sequence published years ago.

Simply evaluating how real the threat of bioterrorism is, and what form it might next take, is fraught with difficulties. The CIA keeps an eye on more than twenty nations it suspects of developing or hiding biological weapons. The CDC classifies biological agents according to their perceived level of risk to national security.[2] Class A, the highest risk, includes anthrax, smallpox, botulism, plague, tularemia, and hemorrhagic fever viruses such as Ebola. All can kill swiftly.

Plague is potentially one of the deadliest. It has caused global pandemics since A.D. 541, when it wiped out 50 to 60 percent of the world's inhabitants. The Great Pestilence or Black Plague of the fourteenth century killed 13 million people in China and went on to kill 20 to 30 million in Europe—one-third of the continent's population. In modern times, plague outbreaks have occurred in China, India, and Madagascar, and sporadic cases have cropped up in the United States. In June 2000, public-health officials detected the plague in rodents in twenty-two counties surrounding Sacramento, California. In 2002 the Donner Memorial State Park in California's Sierra Nevada mountains was closed just prior to Labor Day weekend when local authorities discovered that plague had killed at least one squirrel and had infected a park ranger's cat.

Plague comes in two forms, bubonic and pneumonic. Both are caused by the bacteria *Yersinia pestis.* In bubonic plague, fleas become infected when they bite an infected rodent or other small animal, and transmit the disease when they later bite people. The bacteria make their way from the skin to nearby lymph nodes, where they multiply rapidly, swelling the lymph nodes to as much as four inches across. They then move into the bloodstream and destroy small blood vessels, cutting off the blood supply to extremities like the fingers, toes, and the nose, which develop gangrene and turn black (hence the name Black Plague).

Pneumonic plague is rarer, but can be spread as an aerosol,

sprayed into a crowd. It spreads prolifically; a person merely inhaling droplets expelled when an infected person coughs can contract and subsequently spread the bacteria, too. Within one to six days, victims experience chest pain, a cough, and severe pneumonia. Pneumonic plague is universally fatal if untreated, and even with antibiotics, 57 percent die compared with 14 percent of those with bubonic plague. The Soviet bioweapons program in the 1950s and 1960s developed an aerosol form of plague that was easy to grow and disperse. There is a vaccine, but it is not licensed in the United States.

Tularemia is a bacterium found in rodents and rabbits.[3] About 100 Americans are infected each year, typically after being bitten by infected deer ticks or deer flies. Martha's Vineyard, the resort island just south of Massachusetts's Cape Cod, has had a tularemia problem for twenty years. In 2000, fifteen men were infected and one died. Then ten men there contracted it and one died in 2002.[4] Health officials later determined that eight of the men in the latter outbreak had recently mowed lawns or cut brush, and had probably kicked up the pathogen by disturbing the remains of dead rabbits. Symptoms include sudden fever, chills, headache, muscle aches, joint pain, dry cough, progressive weakness, and pneumonia. The victim's eyes can swell painfully, and he can develop ulcers on the skin or mouth. The organism does not spread from person to person, but if left untreated, it will kill 30 to 60 percent of people who contract it, owing to respiratory failure or shock. Even with treatment, 2 percent die.

Terrorists could spread tularemia by spraying it in aerosol form from small planes.[5] It can survive for weeks in wet straw or other vegetation. A World Health Organization committee reported in 1970 that if released over a city of 5 million, tularemia would incapacitate 250,000 people and kill 19,000. The United States and Japan studied it as a biowarfare agent as far back as World War II, and it was one of several biological weapons the U.S. military stockpiled in the late 1960s and presumably destroyed in 1973. Intelligence reports indicate the Soviet Union continued weapons production of antibiotic-resistant and vaccine-resistant strains into the early 1990s.

Another feared Class A agent is actually a toxin produced by the bacteria *Clostridium botulinum.* This toxin causes botulism.[6] Normally botulism results when the bacteria grows and produces the toxin in improperly canned food that the producer fails to heat thoroughly; heating to 185 degrees Fahrenheit kills the organism. The botulism toxin binds to the ends of nerve cells and prevents them from functioning, causing muscles to become paralyzed. Victims first have trouble swallowing, speaking, or seeing, and may die when paralysis obstructs their breathing—all within two to seventy-two hours after exposure. Doctors can inject an antiserum that neutralizes the toxin, but it is available only through the CDC and state and local health departments. Recovery can take months as the body slowly grows new nerve endings. The Aum Shinrikyo sect in Japan was found to be growing large amounts of *Clostridium botulinum* in 1995, but they were stopped before they had a chance to spread it around.

The hemorrhagic fever viruses kill by causing massive internal bleeding, turning a person's blood vessels into sieves. Potential bioweapons include Ebola, Marburg, Lassa, Rift Valley, Omsk, New World, Kyasanur Forest, and yellow fever viruses.[7] None occurs naturally in the United States; most are limited—for now—to Africa and tropical areas of South America.

A person exposed to any of these viruses whose immune system did not quickly clear it would suffer fever, rashes, body aches, headache, fatigue, and internal bleeding that got rapidly worse. Symptoms appear in two to twenty-one days after exposure. The hemorrhagic fevers are highly infectious, usually transmitted by blood. Terrorists could spread the disease by contaminating the blood supply at a donation center or hospital. Some of the organisms could kill up to 90 percent of people infected in an outbreak. Health-care workers who handle victims infected with these horrific diseases must wear face shields, breathing masks, and full-body gowns, and operate in negative-pressure rooms that don't allow any air inside to escape. There is no approved medication to cure any of the viruses, and there is a vaccine for yellow fever but not the others.

Certainly the trauma that could be inflicted on us by bioterrorism is a grave concern. The anxiety of a potential attack alone imposes a psychological burden. And we can glimpse in the stories of the 2001 anthrax survivors the insidious, long-term impact that even such a limited biological attack can inflict. Although the news had moved on a year after the November 2001 tragedy, the suffering continued for the victims who had been infected and lived. They didn't simply take the antibiotic Cipro and get better. Their ongoing struggles are deeply troubling.

Dave Hose, fifty-nine years old, was six feet eight inches tall and weighed 262 pounds. A postal service supervisor for more than eleven years, he could pick up a bulging, seventy-pound mail sack off the floor with one big hand. Dave had become one of six people who ran the mammoth U.S. State Department Annex 32 mail-sorting facility in Sterling, Virginia, and it was a big responsibility. The wide-open warehouse covered more ground than a football field. Eighty employees operated four of the highest-speed automated sorting lines in the country around the clock, processing five tractor-trailer loads of letters and packages every day. The place handled all the mail going to the U.S. State Department and its 260 foreign posts, including the department's highly sensitive diplomatic pouch service.

Like many federal employees, Dave had taken the September 11 airliner attacks personally. He was as angry as he was sad. His strong emotions seemed to make him work even harder. But on this Monday, October 22, he felt tired his entire shift. On his way home he stopped briefly at Wal-Mart. By the time he returned to his car he had broken out in a cold sweat, and as he exited the parking lot he became dripping wet. The weather wasn't hot, and he didn't have a fever. He was just sweating profusely, and his strange condition continued all night.

The sweating finally ceased during his drive to work the next morning, but it flared up later in the day several times. By afternoon, muscles across his body hurt, and when he got home at five-thirty he began throwing up—blood, not food. He was coughing hard. This new ordeal again lasted the night.

On Wednesday, Dave decided he had better go to Winchester Medical Center. While driving himself there along the suburban roads, he had to shield his face with his hand because the sunlight was burning his eyes badly. That was weird—it was October and it wasn't especially sunny.

The emergency-room nurses paid little attention to Dave during his first two hours there. Finally a doctor saw him and told him, "You probably have the flu."

"No, it's not that," Dave replied.

"Well, you could have pneumonia. Or it could be bronchitis."

"No, no," Dave answered. "I've had them all and I know what they're like. I have never felt anything like this. It's different. I think you should do a blood test."

"For what?" the doctor asked.

"Anthrax."

"Anthrax?" the doctor replied, incredulous.

Dave showed the doctor his State Department ID card. He explained that he worked in the department's primary mail facility. He reminded the doctor about Robert Stevens, the sixty-three-year-old layout editor at American Media in West Palm Beach, Florida, who had died two weeks earlier of multiple organ failure after opening an envelope laced with deadly white powder. The shocking case was all over the TV news. And rumor was going around that several postal workers who sorted mail going to the congressional buildings in Washington, D.C., were in hospitals with anthrax. Dave was anxious about whether he might be the next victim, but the doctor waved him off. So Dave told him one other piece of information he wasn't really supposed to share.

"We've already sent six suspicious letters to the FBI. They looked odd. They had white powder spilling from their corners."

With that, the doctor conceded that they'd better check for anthrax, so he drew blood, sent it to the lab, and gave Dave a shot of Cipro. He also handed Dave a prescription for Cipro pills and told him to get it filled right away.

It was nearly evening by the time Dave's four-hour hospital ordeal was over. He dragged his exhausted, aching body to the local pharmacy. The pharmacist told Dave the prescription would cost $300, but because of the late hour he couldn't get Dave's insurance company on the phone to approve the unusual order. So they naively agreed to let the matter wait till morning. Dave, feeling pain and dread, endured the night with his wife, Connie, by his side. Neither one slept.

Dave's phone rang at 7:55 the next morning, Thursday, October 25. It was the hospital. The physician on the line forcefully said seven words: "You have anthrax. Come in here. Now."

Dave's daughter immediately drove him to the medical center. The emergency-room staff wheeled him in a wheelchair directly to Intensive Care. From the moment the nurses put a gown on him, Dave's world became a medical blur. Before he knew it, he had four intravenous lines in his arms. The nurses drew blood and wrapped a cuff around Dave's big arm. His blood pressure had shot up to an unbelievable 268/136—his circulatory system was a pressurized bomb ready to burst. His heart rate was a frantic 165. And since his arrival a fever had suddenly set in, fluctuating randomly from 102 to 104. Dave felt physically sick, but the shocking news reports he'd seen about anthrax were what really scared him.

The next four days were hell. Dave's body and brain were racked with fatigue, drugs, and blood loss. His face was ashen behind the ever-present oxygen mask. He drifted in and out of consciousness, but he was lucid enough, at times, to hear the doctors talking, and at one point he heard them muttering that they didn't know whether he would live or die. Since he couldn't eat, they infused heavy doses of glucose into his bloodstream, then insulin to counteract the extreme blood-sugar roller coaster that followed. They also infused him with Lasix, a diuretic that forced him to urinate every two hours to help eliminate the toxin. He had to take a cupful of pills four times a day.

The primary physician, Mark Galbraith, was not certain what else to do.[8] "No one is supposed to survive inhalation anthrax," he told his fellow doctors. Of course they would try to counter each symptom, but they knew the chances were high that they'd fail. As the bacteria raged, fluid collected around both of Dave's lungs; doctors pushed a large-bore needle through Dave's chest to drain its interior with a syringe. Five ulcers—holes torn by the bacteria—opened in Dave's stomach and began bleeding badly. Physicians stuck a tube down Dave's mouth and esophagus into his stomach to cauterize them, zapping the internal blood vessels with an electric current. Then Dave's lymph nodes swelled tremendously as the bacteria and immune cells inside them multiplied in an out-of-control fight, searing him with pain such as he had never felt before. When his heart began to beat irregularly from the stress of his body's battle, Galbraith infused more medication and kept an eye on the high-powered defibrillator paddles in case Dave's heart suddenly stopped.

Galbraith soon learned that eight other people in Florida, the Washington, D.C., area, and New Jersey were now in hospitals with inhalation anthrax. Four of them were mail sorters in a District of Columbia sorting facility, and two had already died, one of a heart attack, the other within six hours of being admitted.

The doctors worked over Dave hour by hour, day after day, using every trick they knew to relieve the stress on his heart, stop internal bleeding, and help his big body expel toxin, attempting to buy time for the drugs and his immune cells to mount a growing counterattack against the multiplying pathogen. Remarkably, by Day 7 it seemed the anthrax was beginning to weaken, although the battle continued for eleven more days. During that time another victim, a female hospital worker in New York City, died. Finally, on November 9, Galbraith became convinced that Dave had beaten the enemy, and he sent the patient home. Dave's daughter had to push him into his house in a wheelchair, however, because his body had been so badly weakened he could not walk on his own.

Home offered little solace. Even though the deadly pathogen had not killed Dave, it had torn up his muscles and significantly damaged his heart and liver. A strapping man three weeks earlier, he could not even step across the living room of his brick town-house apartment without collapsing onto the couch. He was taking large doses of two heart medications. For some reason doctors could not discern, he now had terrible asthma, and had to suck on an inhaler frequently. He quickly developed arthritis in his knees, another mystery.

Months later, by July 2002, Dave was still functionally crippled. When he had begun physical therapy after his hospitalization, he had managed to walk for three minutes on a treadmill. Now, seven months later, he could muster only six minutes. A doctor from the CDC who visited Dave intermittently still could not be certain anthrax was cleared from his system. He told Dave his heart and lungs were so damaged he might never return to work. Dave's blood vessels were scarred, meaning they might never heal fully enough to fuel his muscles properly. He was still dependent on drugs. "My body aged twenty years in those sixteen days," Dave said in August 2002. "I am unable to do anything." He also was convinced that anthrax had affected his brain. "My memory is screwed up. That's why I hesitate a lot when I talk." A neuropsychologist he had been seeing told him that seemed strange.

Dave's condition has worn on his mental state, too. Here he is, the big supervisor, sitting at home day after day, month after month, useless, collecting only two-thirds of his pay through workers' compensation. His wife, Connie, has to do everything, and he apologizes to her frequently for being a burden. And who would ever hire him? There is no work he can do. He feels he has no future, and has become seriously depressed. "I am stuck," he says. "And I am sure I am a real big drag." His neuropsychologist has put him on antidepressants.

Dave's acknowledgment that his condition has strained his mind as well as his body might well be behind his belief that he and the others infected with anthrax might have been victims of the U.S. government itself. He explains that he now suspects that the anthrax mailings were part of a secret experiment by the federal government to test how well the pathogen would spread. "Within three hours of the time I was admitted to the hospital," he notes, "an FBI agent was standing at my side. I asked him if this attack was because of foreign terrorism. He said no, it was domestic. How did he know so soon? . . . Nurses took numerous vials of my blood during the first two months; everybody wanted it, constantly—the hospital, the Virginia Department of Health, the CDC, the FBI. . . . I no sooner had gotten home than an FBI agent told me the anthrax was of the Ames variety—but scientists didn't 'decide' this was what was showing up in lab tests until later. How did he know? I've since learned that I was exposed to the latest innovation—it was the smallest, most modified, highest statically charged anthrax that had ever been seen."

Dr. Galbraith, who was still caring for Dave in September 2002, was frustrated by his patient's deteriorating physical and mental condition. He felt somewhat helpless. There had been only eighteen U.S. cases of inhalation anthrax in the past century, the most recent twenty-six years ago, in 1976. No one had ever survived. "When we did finally stabilize him," Galbraith says of Dave, "we wondered, 'Okay, what do we do now?' There is no medical literature to follow. We have no cases to refer to." Galbraith and the doctors treating the other survivors continue to wing it. "The arthritis in his knees, that could be from age," Galbraith ponders. "The asthma—that's a real puzzle; he does do better with the medication. His muscle pain and fatigue? That doesn't make sense. His heart rate doesn't shoot up during physical therapy, his blood oxygen is not saturated, yet he says he can't take another step. Maybe the toxin has damaged him in some way we don't understand." In December 2002 five blood clots suddenly formed in Dave's lungs, putting him in the hospital again for more than a week.

Galbraith also wonders if the toxin has damaged Dave's brain. "Or maybe he is suffering from post-traumatic stress disorder—a rare victim of an unprecedented terrorist attack." Dave doesn't know about that. All he knows is that he feels useless, he is a burden, he is not the same person he was, yet he sits in his house all day, supposedly lucky to be a survivor. "Why me?" he asks aloud in frustration. "Why do I have a disease no one knows how to treat?"

Anthrax is transmitted as a spore, a minute kernel of protein and genes. The hardened speck can survive high heat, dryness, and extreme cold for years. Only high-pressure steam, fire, radiation, or caustic chemicals can destroy it. Although it seems lifeless, the spore is actually a dormant bacterium. Like a seed, it will remain silent until it lands in the skin, gut, or lungs of a living being. There it will germinate and reproduce furiously.

Anthrax is uncommon in nature, yet isolated spores do find

their way into soil, and grazing sheep or cows may occasionally breathe one into their snouts as they snuffle the ground, pulling up grass.[9] If enough spores germinate, the animal will die quickly. A person who handles its carcass might absorb some spores in his chapped skin—a condition that can lead to cutaneous anthrax. Gastrointestinal anthrax can occur at odd times, too. In July 2000, a Minnesota farm family slaughtered a cow they did not know was sick.[10] They ate the steaks and suffered cramps, fever, and diarrhea. Health officials tested another of their cows that died mysteriously and found anthrax, and started the family on antibiotics, which fortunately cured them.

In still rarer but much more lethal cases of inhalation anthrax, a person might inhale spores that have become airborne. The person may clear the large spores by coughing, but the tiniest spores, 1,000 times smaller than the head of a pin, can penetrate the alveoli, the little air sacs in the lungs. The body's immune cells engulf the spores, and that's when they transform into bacteria that reproduce by the millions in a single day. The new bacteria pour into the bloodstream, producing prodigious amounts of toxin, sending the victim into a tailspin like the one Dave suffered.

Anthrax is insidious. Symptoms may develop within two days of exposure, but can arise as many as eight weeks later if the spores take time to establish themselves in the lungs. Initially, the ill-fated human host will feel as if he has a cold, but once the person begins to cough or feel feverish, he will usually die in one to three days—the infection is already raging, taking over the chest and putting the victim into fatal septic shock from the poison overload. Without treatment, almost 100 percent of victims die. When a person dies, so does his immune system, and the bacteria consume the body. Once the feeding orgy ends, they revert into spores and wait for the next host.

Clearly, even though Dave Hose took massive doses of Cipro, anthrax still ravaged his body. The other victims who survived fared no better.[11] A year after their trauma, most still suffered from persistent fatigue and memory loss. Only one was able to return to work; most of the rest continued on disability. Many were taking antidepressants and had regular sessions with psychiatrists. Some were prone to fits of intense rage; others experienced ongoing chest pains.

One of the glaring lessons the country learned from the anthrax attacks is that developing and manufacturing sufficient quantities of drugs to stop a terrorist-spread epidemic presents a daunting challenge. When the anthrax-laced letters began arriving, doctors gave congressmen, journalists, and postal workers Cipro as a preventative. A vaccine was developed thirty years ago, but it was only approved for military use; defense officials gave it to troops during the Gulf War because they thought Saddam Hussein might use anthrax as a weapon. Even if the vaccine were approved for civilians, by the time doctors detected anthrax infection in each of the October 2001 victims, it would have been too late.

Cipro, the abbreviated brand name of the antibiotic ciprofloxacin, is made by just one supplier: Bayer AG of Germany. Furthermore, ciprofloxacin kills the anthrax bacteria but doesn't neutralize their toxins, so it must be taken within two days of exposure—better yet, before exposure occurs. Dave's inability to fill his prescription could have allowed his death had he not been given the Cipro injection at the hospital that day. The rush on Cipro quickly outstripped Bayer's ability to ramp up production, as well as the public-health system's ability to distribute and administer it. When people across the country called their doctors in panic, demanding preventive doses, the drug simply wasn't available in many places.

Cipro is far from an ideal antidote, either. It can cause extensive side effects, like fatigue, nausea, dizziness, joint pain, and tendon rupture, along with agitation, nightmares, and even paranoia, which could have contributed to Dave's conspiracy theory. It can dangerously amplify the action of certain other drugs, such as blood thinners. Yet the FDA recommended Cipro during the scare because it thought the anthrax used might have been engineered to resist the two more-common antibiotics that could fight it: penicillin and doxycycline. After mid-October, when the CDC's analysis of the anthrax showed it was susceptible to doxycycline after all, the FDA recommended that drug as well. If a mass outbreak had followed, though, even the more plentiful penicillin and doxycycline would have posed problems. About 1 percent of Americans are allergic to penicillin; the reaction itself can cause death. And doxycycline can cause dangerous sun-sensitivity. The lesson was clear: The United States had neither a sufficient supply of vaccine to fend off a terrorist anthrax attack, nor drugs that could cure it without causing their own widespread complications.

While the anthrax scare put bioterrorism on the public map, the medical community was much more alarmed by a different pathogen, the dreaded old scourge smallpox. For one thing, as the media reported, it takes considerable expertise to produce "weapons-grade" anthrax. A high concentration of just the spore form of the organism is required, and that must be treated so that the particles are very fine and will stay in the air a long time. The anthrax in the letter to Senator Leahy had traces of silicon on it, suggesting it had indeed been specially treated to keep it from clumping. The difficulty of concocting lethal anthrax is one fact that led investigators to suspect that the pathogen had come from a U.S. government research lab. As of October 2002, investigators still had not definitively tracked down the lab from whence the anthrax came—or if they had, they weren't saying. But clues pointed to the U.S. Army Medical Research Institute of Infectious Diseases—the military's top infectious disease group at Fort Detrick, Maryland, known by its aggressive-sounding acronym, USAMRIID—where such high grades are stored.[12]

The greater saving grace about anthrax is that it is not contagious; one person cannot pass it to another. A twisted individual has to mail it to you, blow it into the ventilation system in your office building, or spray it inside your train car or bus, and then you have to inhale it. To kill many people, scores of terrorists would have to send thousands of letters or infiltrate numerous buildings or buses, day after day.

The story is very different, and much scarier, regarding smallpox. This disease, caused by the variola virus, is both extremely contagious and very easy to spread.[13] That's why, as soon as the word of the anthrax mailings hit, scientists at the CDC, the CIA, and the Pentagon immediately worried that a smallpox attack might be next.

Smallpox is so feared because a maniacal group could spark an epidemic that infects hundreds, possibly thousands of people. In June 2001, before the anthrax scare, defense experts at Andrews Air Force Base outside Washington, D.C., ran a computer-driven exercise called "Dark Winter," which dramatically enacted the chaos a terrorist release of smallpox would create.[14] Fourteen senior government officials played the roles of members of the National Security Council (NSC), and several members of the press represented the media. For thirteen days they reacted to news of the outbreak and initiated simulations of real-world responses they would take to contain its spread. The virtual war game began with a first smallpox victim, diagnosed in Oklahoma City, of whom the NSC was informed. No one knew the source of the infection, but other victims soon appeared in Georgia and Pennsylvania. By the time the attack was discovered, the original victims had already traveled and infected others. The model assumed 1,000 people were initially infected by aerosal clouds released in three states. Within two months the contagion had spread to twenty-five states and fifteen countries. The government quickly ran out of vaccine. Three million people were infected, 1 million people died, and interstate commerce stopped. The government had still not contained the epidemic when the exercise ended. Fortunately, the vaccine supply, which is our main defense against smallpox, is now much larger than was known at the time of the Dark Winter simulation. But an outbreak would still be a catastrophe.

Dark Winter was modeled on a real outbreak in Yugoslavia in 1972. According to internal CDC documents, a thirty-five-year-old Yugoslavian man who had unknowingly become exposed had returned to his home in Kosovo after a pilgrimage to Mecca. Shortly following his return he felt achy, but never got really sick, perhaps because he had been vaccinated against smallpox during childhood. However, he infected another man in a local market who was not so lucky. His initial smallpox rash went undiagnosed, and he was treated with penicillin. But when he developed a high fever, he took the bus to a local hospital, where he became gravely ill. Doctors sent him by ambulance to the Dermatologic Hospital in Belgrade. Staff there misdiagnosed his rash and accompanying fever as a bad reaction to the penicillin. But when the whites of his eyes turned a bloody black, and lesions broke out across his body, he was transferred to a second, more general Belgrade hospital, where he died two days later. No one along the way realized the man had smallpox.

It took about one month and twenty cases before anyone realized they were dealing with a smallpox outbreak. Follow-up research showed the man who died had infected thirty-eight people, all but two of them patients or staff at the hospitals. The initial pilgrim had infected ten other people as well. They in turn spread the disease widely. Most of the population had been vaccinated, but for many it had been long ago, leaving them with weak or no immunity. Yugoslavia's authoritarian Communist leader, Josip Tito, declared martial law, isolating patients and their contacts and quarantining the affected regions. Luckily, the government had 18 million doses of vaccine available, and residents of Kosovo and Serbia, the two provinces involved in the outbreak, were vaccinated. In all, 175 people contracted smallpox and thirty-five died. People who had never been vaccinated before they became infected were three to four times as likely to die as those who had been vaccinated earlier.

The U.S. death toll in Dark Winter was much worse for several reasons, according to a follow-up report by the U.S. Center for Strategic and International Studies, a lead player in the war game. Virtually no American has immunity anymore; vaccine protection fades after a decade or two, and vaccinations stopped thirty years ago. Very little vaccine has been stockpiled. There is no "surge capability" in the hospital and public-health systems to care for so many victims, and no surge capability in the pharmaceutical industry to crank up vaccine production. Conflicting roles of federal and state agencies and of the Defense Department, and complications concerning the civil liberties of people associated with quarantine and isolation, further hampered the prevention efforts taken during Dark Winter. Major rifts became apparent among federal, state, and local governments, and between government and the private sector, which compromised the nation's ability to limit loss of life, suffering, and economic damage.

Despite the crash program set up by the CDC after the anthrax scare to dramatically increase production of smallpox vaccine, we must still be on high alert about the dangers of this horrible pathogen. The official word is that, today, the smallpox virus exists only in freezers at two laboratories: CDC headquarters in Atlanta, and a Russian virology institute known as Vector Laboratories, outside of Novosibirsk, Siberia. But throughout the 1990s, American intelligence agents found increasing evidence that stores of smallpox were secretly stashed at germ warfare labs in Russia, North Korea, Iraq, and even France. They also had indications that other countries, including Iran, Pakistan, India, and China, either already had smallpox or were trying to obtain it. Informants told them terrorist organizations were also in the market, including those run by Osama bin Laden and by the Aum Shinrikyo extremist sect of Japan; the latter group used umbrellas to pop bags of sarin nerve gas in a Tokyo subway in 1995, killing twelve and injuring 5,500. Whistle-blowers said Russia and North Korea were designing and even testing missile warheads that could fly hundreds of miles and then explode, raining smallpox down on the population below. In 1998 a Russian defector, Ken Alibek, who had served as deputy director of the Soviet biological weapons program, spoke freely about such activities, noting that the Soviet war machine had built facilities capable of producing tons of smallpox concoctions annually.

We should take the smallpox threat seriously because it is truly a heinous scourge. Epidemiologists say it has killed more human beings than any other infectious disease: 300 million to 500 million people in the past millennium, by various accounts. People can spread the virus simply by talking, kissing, or shaking hands. If you inhale a single particle, it can multiply stealthily in your healthy cells and spread throughout your bloodstream. Ten days later you will have a high fever, a headache, and muscle aches. Ulcers will begin to form in your mouth, throat, and lungs. The virus will accumulate beneath the deepest layer of skin, from your head to your feet. Then it will grow outward. A red rash will appear all over your body. Within days it will turn into blisters, then hard, itchy pustules plastering your body. In the worst cases the pustules will join together, raising whole regions of skin from the muscle beneath. The skin feels as if it were on fire, the pain so intense victims cannot bear to move.

As horrible as the blistering is on the outside, the destruction inside is worse. Like all viruses, smallpox particles burrow into cells to reproduce. They penetrate lymph nodes, the spleen, and bone marrow. Scientists are not certain, but they think the new copies destroy cells as they exit, each time infecting and destroying many subsequent cells. The immune system attacks healthy cells containing the virus in an effort to limit the virus's replication. So while the enemy tears up our flesh, we bomb it, too. The combined onslaught destroys a person's organs, and they die gasping for breath during lung failure, heart attack, or toxic shock. Three of every ten people who become infected with smallpox die. If you survive, the pustules will scab and fall off, leaving you scarred for life.

There is no effective treatment for smallpox. It's a battle to the death: you or the virus. Your fate lies in a race between the virus trying to replicate and infect more cells, and your immune system trying to ramp up enough to kill it off. Young children and elderly people are most at risk, because their immune systems are underdeveloped or waning.

Smallpox has wreaked havoc since it swept across Egypt 3,000 years ago. In the early sixteenth century the Spanish brought the highly contagious disease to the Aztec people, decimating the population in the Americas by up to 90 percent in fifty years. Half of colonial Boston's 12,000 inhabitants were stricken in 1721. During the French and Indian Wars of the mid-1700s, the British genocidally distributed smallpox-laced blankets to Indian tribes as gifts. But Edward Jenner's development of a smallpox vaccine in 1796 started mankind's long march to gain the upper hand. Although 48,000 Americans caught the contagion and one-third of that number died in a 1930–31 epidemic, the last U.S. victim was in 1949. Better vaccines and nationwide vaccination programs spread across many countries, stamping out the pathogen. The United States stopped vaccinating the public in 1972. After a final, aggressive, decade-long vaccination assault by WHO, the international group declared smallpox "eradicated" worldwide in 1979. A Somalian man was the planet's last natural case, in 1977.

One result of this unprecedented success in disease control, however, was that the government had little smallpox vaccine on hand in the wake of the autumn 2001 terrorist attacks. WHO once had 10 million doses stored in Geneva, Switzerland. But in 1990 it felt the pathogen was no longer a threat, so it decided to destroy most of the vaccine. Workers in full suits of protective gear incinerated all but a half-million vials, leaving one dose for every 12,000 people on earth. What a terrible irony it would be if our eradication of smallpox is turned against us through a biological attack.

Some defense experts say the Dark Winter scenario for the terrorist spread of smallpox is not likely because even if terrorists got their hands on smuggled vials of virus, they would have difficulty safely working with the microbes and would likely die themselves. But as the September 11 plane crashes into the World Trade Center towers proved, some extremists are not afraid to commit suicide for their cause. Because smallpox is so contagious, a single "smallpox martyr" could inhale the stolen virus, wait a week until he is highly contagious, then begin coughing on people standing four deep waiting for red lights to change at the busy street corners of Oklahoma City during lunchtime. Or he could stand shoulder to shoulder with commuters in a crowded subway car during rush hour in New York City or London or Tokyo, coughing purposely on the dozens of people entering and exiting the car at every subway station. Those victims would then become highly contagious themselves, passing the killer virus to their husbands or wives or children with an innocent goodnight kiss, or perhaps sneezing into their palms, then shaking their coworkers' hands. Ten people might infect ten others, and they ten others, exponentially spreading the killer to a hundred people, then a thousand, and so on if left unchecked.

However, now that the public is alerted to the possibility of a smallpox attack, and plans and vaccine stocks are in place for a massive vaccination campaign, it is unlikely that any one martyr could infect numerous people before the public-health system sprang into action. More worrisome, but more difficult to pull off, would be the coordinated release of an aerosol of smallpox. Once again, when the first victim showed up at a hospital or doctor's office, a rapid public-health response would likely keep the disease from spreading much beyond the initial group of victims.

It's also not farfetched that smallpox could be smuggled out of defense labs. In January 2002, newspapers broke the story that a secret, internal U.S. Army inquiry indicated that lab specimens of anthrax spores, Ebola virus, and other pathogens had disappeared from USAMRIID's biological warfare research facility in the early 1990s, during a period of labor complaints and tension among rival scientists there. Defense officials said it was unclear whether there had been any actual theft; the discrepancies could have been due to misplaced samples or clerical errors. But the incidents and their cover-up suggest lax procedures that undermine confidence about the security at top defense labs. USAMRIID is also suspected as the source of the anthrax used in the October 2001 attacks, but more than a year later no one had yet been charged with a crime; the FBI investigated former USAMRIID scientist Stephen Hatfill in a very public way, condemning him yet never charging him—some thought just to show the FBI was doing something.

The lack of certainty about who might already have their hands on smallpox is one reason the production of large quantities of new vaccine is so vital. There is no way to know if people vaccinated against smallpox decades ago are still immune today. What little research there is indicates that antibody levels in vaccinated persons decline substantially over time, providing only 85 percent protection after ten years, and just 50 percent protection after twenty years. And since the last American exposure was in 1949, few people have natural immunity anymore. Essentially the entire American population is susceptible.

This is doubly woeful because the vaccine not only prevents a person from contracting the disease, but can prevent it even after infection if the victim gets an injection within four days after exposure. The vaccine can actually contain an epidemic.

That's why one of the first significant steps taken in the war on terrorism was to initiate a massive buildup of new smallpox vaccine. In some ways, the events that unfolded were an inspiring example of how effectively we can meet the germ threat if we are sufficiently mobilized. But the saga also highlights the troubling limitations of our privatized vaccine system—and that we must be devoting more resources to the development of vaccines if we are to be truly prepared.

On November 28, 2001, Tommy Thompson, U.S. Secretary of Health and Human Services, awarded a single smallpox vaccine contract—a $428 million order—to Acambis, a little British biotech firm with operations in Cambridge, England, and Cambridge, Massachusetts. Acambis had only 160 employees, and had lost $16 million the year before, but it was the only company in the world prepared to make the product.

A year later, Acambis was well on its way to accomplishing its goals. It had a big head start. For a decade the company had tried to develop vaccines against exotic jungle diseases such as yellow fever, dengue, and typhoid. In 1999 Acambis answered a U.S. government ad in a trade magazine. Defense officials analyzing the bioterrorist threat were increasingly worried that smallpox could be a terrorist's ultimate weapon, and the CDC had decided to look for a company to develop a new smallpox vaccine and produce and maintain a stockpile of 40 million doses. In September 2000, Acambis won the contract, worth $343 million, whereby it would maintain the stockpile for twenty years as well as build new production facilities. The company was slated to begin delivering vials by 2004.

Under Thompson's deal, Acambis had to produce an additional 155 million doses and deliver them to the CDC by autumn 2002, an incredibly accelerated timetable. The company had to increase the initial lot from 40 million to 54 million, and deliver those before the end of 2002 as well. The vaccine also had to be tested. The FDA would try to speed approvals for clinical trials. If all went according to the wildly ambitious plan, an inoculation program could be instituted in 2003.

Meanwhile, researchers at the National Institute of Allergy and

Infectious Diseases (NIAID), a branch of the National Institutes of Health in Bethesda, Maryland, assured Thompson the existing federal stockpile of 15 million doses could be stretched to 77 million doses; the old vaccine was highly concentrated, and vials could be diluted to create more vaccinations. All told, Thompson figured he'd get 209 million doses from Acambis plus the extended 77-million stockpile dose, for a total of 286 million inoculations—one for every man, woman, and child in America.

As of February 2003 the work was moving along. The early human trials were promising. The pressure had been reduced, too, because Aventis Pasteur, headquartered in Swiftwater, Pennsylvania, had announced it would donate an additional 85 million doses that had been in its freezers for forty years; the company and government had kept quiet until then so they could check to see if the vaccine was still viable. In October NIAID announced that not only were the inoculations safe, but diluted samples had been tried on more than 100 volunteers and worked fine. The cache could be stretched to 300 million doses.

Acambis's plan is a model of how the American economic machine can kick into gear if we mobilize it with money. And the company's readiness to respond is testimony to the vision of Tom Monath,[15] one of the vanguards we are so fortunate to have in the war against pathogens. The sudden ramp-up meant a dramatic change in life for Monath, Acambis's vice president of research and medical affairs. Monath had been the scientific driver of Acambis since its founding in 1997 in a merger of two predecessor firms. He had a medical degree from Harvard. In the 1970s he beat paths through the African jungle, where he led research into yellow fever and discovered that mice were responsible for spreading among humans Lassa fever, a vile hemorrhagic virus that turned people's organs into bloody mush in little more than a week. He went on to become director of the CDC's viral diseases lab in Fort Collins, Colorado, and the chief virologist at USAMRIID.

A straight talker whose short gait, erect posture, and staccato speech reflect his quasi-military assignments, Monath had put Acambis on its research course to develop vaccines for West Nile virus, dengue fever, and Japanese encephalitis. He had to immediately transform Acambis into a manufacturer capable of churning out thousands of gallons of vaccine in short order. Seated squarely at his wooden desk in Cambridge, Massachusetts, he reluctantly pushed the *Physicians' Desk Reference* to one cluttered corner and placed a bound report front and center, the one labeled "Vaccine National Stockpile for Containing the Threat of Bioterrorism."

Monath's first step was to subcontract Baxter BioScience, the biotech division of Baxter Healthcare Corp., to grow volumes of virus at an undisclosed European plant. That would buy time to expand a small, dormant production facility Acambis had begun to refurbish in Canton, Massachusetts, just south of Boston, where the company could purify the virus, blend it with a carrier fluid, and fill vials with a live, though attenuated, vaccine. Acambis would have to convert the plant into a big, high-tech, smallpox vaccine factory. Monath hired armed guards to patrol the 47,000-square-foot brick building, innocently set in a landscaped industrial park in the middle of suburban Canton.

The energetic Monath handled the ongoing corporate research chores from Acambis's offices in a brand-new building on Sydney Street, a half hour north in Cambridge. It was a clever place to be, anonymously set alongside other similar buildings, with the Massachusetts Institute of Technology bordering its backside. Casual observers might not notice that the straight street in front of the green glass building was devoid of traffic lights and parking spots, providing no reason for passing cars to slow down. Certain stocky men slyly strolled the sidewalk, sidling up to people who stopped in front of Acambis's unmarked front door. They were guards, revealed only by shirts monogrammed "Security" that were mostly hidden beneath plain jackets. The company was now the world's lone maker of smallpox vaccine, clearly a potential terrorist target.

Up in his office, Monath was frank with the rare visitor who was allowed in. "Dark Winter is very epidemiologically possible," he says. "I've been involved in biowarfare for years, and I feel good that I'm now able to actually do something to improve national security." He had seen firsthand how rabid viruses like Hanta and Lassa, also on the CDC's list of possible bioterrorist weapons, could rage through entire communities in days, ripping people's insides to shreds. He also had no doubt smallpox was in the hands of the North Koreans and would-be terrorists. "There has been a huge brain drain from the former Soviet Union. At one time 50,000 people in that country were working on biological weapons. Many of them have left since the USSR dissolved. All they had to do was put a vial in their pocket. . . ."

But Monath would not discuss how Acambis would make so much vaccine so quickly, where it would be stockpiled, or how that storage facility would be protected. "We're not supposed to talk about those things."

There is reason to hope, however, that military and civilian health agencies are talking a great deal behind the scenes about how to accelerate development of the other vaccines we need in our arsenal. For years they have separately devised vaccines against biological warfare agents, and the bifurcation has hurt both pursuits. Although the smallpox vaccine was readied with remarkable efficiency, the same cannot be said for potions against the other Class A biowarfare agents on the CDC's list. In a report released to Congress in the summer of 2001, the U.S. Defense Department, which has sole responsibility for protecting troops against biowarfare agents but no responsibility for the public, admitted that its own effort to develop vaccines through commercial contractors was "insufficient and will fail." For example, the program to produce an anthrax vaccine at BioPort in Lansing, Michigan, has been beset with production problems, and some federal officials have called for the military to take over production directly. DoD has been reluctant to support that approach because the agency is not experienced in manufacturing, and worries that its involvement would delay production still further. It has since hired an independent group of experts to advise it on how best to proceed.

The private effort had been altogether uncoordinated until the anthrax attacks. For years, civilian vaccine development had taken place at dispersed university, corporate, and government labs, much of it with funding from NIAID, the National Institutes of Health, or the Defense Department. But the huge bioterrorism cash infusion following the anthrax attacks could help centralize the work. NIAID received $1.2 billion for bioterror-related research, including new vaccines for smallpox and tularemia and testing of experimental Ebola and anthrax vaccines that several institutions were readying for human trials. Work on vaccines for the other Class A agents on the CDC's bioterrorism list was not as advanced but had begun.

Even if we can step up production of needed vaccines, however, the case of smallpox shows that tricky issues will still complicate our defense. Distributing smallpox vaccine nationwide would provide a strong deterrent to terrorists; there's little point in releasing smallpox if most of the country has been vaccinated. But widespread vaccination would cause problems because, as vaccines go, the smallpox potion is more dangerous than most—a problem we might well face with vaccines for other diseases bioterrorists might spread. The smallpox vaccine causes serious side effects in one of every 13,000 people, from severe rashes to brain inflammation. About one in a million people die.

The vaccine is dangerous because it contains a live virus called vaccinia, which is closely related to the smallpox virus, variola. Vaccinia produces smallpoxlike proteins in a person's bloodstream, which prompt immune cells to generate a response that protects against variola as well as vaccinia. Vaccines that use live viruses in this way are common and desirable because they replicate just enough to thoroughly challenge the immune system without seriously threatening it. Vaccines using killed organisms or pieces from an organism simply infuse proteins at the injection site, where they exist for a very limited period; they are safer but elicit a weaker immune response than live vaccines.

A live vaccine carries a greater risk, however, because the virus itself can multiply widely and do damage if the recipient's immune system cannot keep it in check. In 1984, military doctors gave a recruit multiple vaccines when he started training, including one against smallpox.[16] Unfortunately, they didn't know that the recruit had the HIV virus, which had significantly weakened his immune system. The vaccinia virus replicated unchallenged and caused an ugly rash, along with fever, headache, and neck stiffness. The doctors had to treat the recruit for twelve weeks before his condition improved.

The smallpox vaccine can make people with skin disorders like eczema deathly ill, and cause a skin condition equivalent to third-degree burns all over their bodies; millions of Americans have eczema. And for the millions more Americans who have AIDS, or are being treated for cancer with radiation or chemotherapy, and the multitudes of others on every continent, a smallpox vaccination could overwhelm their already weakened immune systems and kill them. Their immune systems are not robust enough to control the vaccine.

Such risks are not worth taking if the chances for attack are slim. What kind of vaccination program to pursue is a complicated question that depends in part on intelligence information. For months after the anthrax attacks, and well into Acambis's production of smallpox vaccine, public-health and defense officials were still debating whether military personnel, health-care workers, or the general population should be immunized at all. The threat of attack, some argued, is only theoretical, and the threat of serious side effects is real. After much consideration, in September 2002 the CDC released a report providing vaccination guidelines to the fifty states, as President George Bush pushed a possible attack on Iraq. The guidelines call for each state to be prepared to vaccinate 1 million people within ten days of an announcement of a smallpox attack. The report stated that 280 million doses could be shipped nationwide in five to seven days. Vaccinations would be voluntary for the public. But the plan sidestepped the vexing question of whether health-care and emergency workers should be vaccinated before an attack; if they were not, they would be standing in line for their own vaccinations during an outbreak, instead of administering vaccinations to others. Yet if they were vaccinated ahead of time, some of them would unnecessarily become seriously ill or die.

Finally, in December 2002, President Bush decreed that 500,000 key health workers nationwide would be vaccinated in the coming weeks. Later, 10 million more health workers, firefighters, police, and emergency medical personnel would be vaccinated—condemning some people to serious illness or death. In January 2003 a panel of experts convened under the CDC criticized the move as hasty and dangerous, and politically motivated. By March, only 12,404 health workers had been vaccinated. Many declined the vaccine because of safety and liability issues.

A far better solution would be an effective vaccine that did not depend on a live virus. Microbiologists are experimenting with that approach, but the work will take considerable time. They have long thought they could succeed by inserting smallpox genes into a far safer "carrier" virus. Injecting that carrier virus into a person's body would stimulate an immune response to smallpox, creating immunity to the disease. Better yet would be a method scientists achieved in novel experiments done in 1992, in which they extracted genetic DNA from an influenza virus, created more copies, and injected just this "naked DNA" into the muscle of lab animals. Much to everyone's surprise, early results indicated that the DNA alone was enough to stimulate an immune response in certain immune cells. Vaccines made with naked DNA would be safer because no carrier virus is used, and therefore there is nothing that could replicate to threaten a person with a weakened immune system. And if effective vaccines of this kind could be developed, they could simply be administered orally or through a nasal spray, faster and easier than shots.

Many labs are now working on new vaccines using this approach, though not yet for smallpox, but no compound has gone to human trials because of one huge worry that remains unresolved: If the vaccine's DNA could make its way into the natural DNA inside a person's cells, it could disrupt the normal genes and potentially wreak havoc. The greatest danger is that by upsetting the DNA regulation of cells, cancer could develop. That has not occurred in any of the many lab animals tested so far, but it is too early to rule out the possibility.

The striking advances being made in genetic research are surely among our most promising tools against bioterrorism. But the same techniques could also be used against us.

Microbiologists could, for example, insert genes into pathogens so the organisms would be unfazed by our drugs, or genetically engineer whole new agents that would be even worse than those that already exist. Ken Alibek, the Soviet bioweapons defector, wrote in his book *Biohazard* that the Soviet Union had engineered versions of the anthrax and plague bacteria that were resistant to antibiotics and even the Soviet nation's own vaccine.[17]

In 1996 microbiologists at the Australian National University in Canberra created a killer while they were trying to engineer a virus that would pass on infertility genes to rabbits, as a way to reduce the rabbit population that was overrunning Australia and causing considerable crop damage.[18] The scientists inserted an infertility gene into a myxoma virus, a relative of the smallpox virus, which would spread the gene across rabbit populations, reducing their subsequent reproduction. But instead of acting like a carrier virus, the myxoma ran amok, killing more than 100 million rabbits.

Startling news came in July 2002, when, as mentioned earlier, SUNY Stony Brook scientists announced they had synthesized a polio virus that was virtually identical to the naturally occurring organism.[19] The Pentagon had sponsored the work as part of research into bioterrorism, to see if a killer organism could indeed be made from scratch in a lab. Polio would not make a good weapon, because most Americans are vaccinated against it and fewer than 1 percent of people infected develop anything more severe than intestinal problems. But it was a place to start, because the virus has a smaller, simpler genome than the more likely biological weapons such as smallpox. The work showed how fast biotechnology is progressing—for potential evil as well as for good. We can no longer assume that any organism will ever be eradicated; even if polio or smallpox was indeed "wiped out," biotechnologists could build it again—and as their techniques improve, they will be able to make the pathogens even more deadly. The world faces a possible biological arms race—an ever-escalating game of increasingly sophisticated engineered attacks, countermeasures, and counterattacks.

Certain defense experts argue that the threat of engineered bioterrorism has been exaggerated. They say the exercise is so scientifically challenging and costly that it could not be readily pursued by terrorists. That might be true for rogue operators, but rogue nations might well have the resources to try. That's why we've heard so much about Saddam Hussein's alleged development of biological weapons. Heads of state like Hussein could also decide to supply the deadly agents to outside terrorist groups.

Perhaps a more persuasive argument about why synthetic organisms won't be used by terrorists is that there is no need for such bizarre creations. As Monath says, "It's hard to imagine terrorists would bother to engineer a new agent. The natural ones are already so deadly." It also would be simpler to add drug-resistance genes to well-known pathogens than to devise new organisms.

Even the use of natural pathogens could prove infeasible, however. Growing infectious diseases in a lab may not be difficult, but processing massive quantities of lethal pathogens, and deploying them in ways that inflict large numbers of casualties, requires huge amounts of money, intricate materials, and rare know-how. The challenge was illustrated late in 2001, when rumors flew that terrorists were trying to rent crop-duster aircraft to spray anthrax over towns or perhaps a crowded U.S. stadium. Crop-dusting equipment would have to be modified extensively to produce a mist of microscopic particles just the right size to penetrate deep into human lungs. International United Nations inspectors who combed Iraq after the Persian Gulf War in 1991 said Saddam Hussein's war machine, perhaps the best-financed rogue operation in the world, had tried to make such modifications, but had failed.

Surely one of the most demanding efforts in the war on terrorism will be developing effective, plentiful antidotes to fight the range of pathogens terrorists might employ. Progress on this front has so far been slow. Microbiologists at a small number of labs in the United States and Europe are urgently investigating new compounds that could thwart pathogens like anthrax and smallpox, saving people who had never been vaccinated, did not get vaccinated in time after an outbreak began, or should not be vaccinated because the inoculation presents a significant risk. A 1999 report from the National Academy of Sciences' Institute of Medicine declared that the best reason to retain the CDC stockpile of smallpox virus is precisely so it can be used to develop new drugs that can beat the disease.

Some scientists working on treatments in government labs try to keep their research secret, to foil enemies who might try to discover what they are up to. But we do know that researchers at several universities and institutes are looking for ways to counter anthrax. A Harvard team led by R. John Collier, professor of microbiology and molecular genetics, succeeded in 2001 in engineering a mutant anthrax protein that attaches to healthy cells and prevents the anthrax toxin from penetrating them.[20] Other researchers are trying to figure out how to stop anthrax from producing toxin in the first place.

We are nowhere near curing smallpox, however. Virtually all research on the disease ended when it was eradicated thirty years ago, with the possible exception of work done in secret by the army. And most modern microbiology and genetic techniques have been developed since that time. As a result, little is published or even known about how the virus works at the molecular level. Furthermore, smallpox infects only humans and monkeys, so lab animals such as mice, normally used widely in disease research, offer no help.

The challenge is greater still. Because smallpox is so incredibly contagious, scientists can study the virus only in labs that can insure the highest level of containment—biosafety level 4 (BSL-4). In these labs, scientists must wear protective "space suits" and breathe through respirators. A closed ventilation system sweeps and filters the air under negative pressure—no air can escape the room even if its specially sealed doors and air locks are left open. At the time of the anthrax attacks, there were only four BSL-4 labs in the country, including one at the CDC and one at USAMRIID. The lack of BSL-4 lab space severely limits who can grow smallpox and therefore develop treatments. Immediately after the anthrax alarm, the CDC and NIH fast-tracked plans to build more BSL-4 labs. But more facilities may be slow in coming, in part because there are few architects and builders who know how to construct them. On a positive note, the CDC held a large conference in January 2002 to bring scientists together with lab builders in a hurried attempt to form a community able to improve the country's biosafety lab infrastructure.

Until that larger research structure is in place, further insight into how our immune system and drugs might interact with smallpox is likely to come from research with related viruses. Smallpox and its kin, monkeypox, belong to the orthopox family of viruses, which has infiltrated rodents, cows, horses, buffalo, camels, monkeys, and apes as well as humans. Brick-shaped and surrounded by a durable coat of fats and proteins, the orthopox viruses are the largest of all animal viruses. They use all sorts of tricks to avoid being destroyed by the immune system. Researchers have made some progress in devising drugs to combat them, which could provide important clues to compounds that fight smallpox itself. One potential drug that has received attention is cidofovir, initially designed to counteract a virus that can cause blindness in people with HIV. Cidofovir inhibits an enzyme that smallpox needs to reproduce once it has slipped inside a healthy cell.[21] John Huggins at Tom Monath's former facility, USAMRIID, has found that the drug works against monkeypox. However, his research also shows that cidofovir can harm the kidneys, and causes other risks associated with nucleotide analogues, the class of compound it belongs to. Much more work must be done before safe antidotes for humans can be prescribed.

In addition to developing a powerful drug defense, we must greatly improve our ability to gather good intelligence about which hostile groups or countries are developing bioterrorism programs. Technology to provide early warning that a pathogen has been released would speed detection of attacks, saving many lives. Once a year the Defense Department's Program Office for Biological Defense invites detection experts to the Dugway Proving Grounds, a salt flat in the Utah desert, to test their latest high-tech versions of the coal miner's canary. Researchers post their devices in the hot white sand and retreat to a trailer a half mile away. Defense officials then release an aerosol cloud of bacteria near the sensors, which simulates a biological agent. After the cloud disperses, everyone rushes back to see if any of the diagnostic contraptions has determined the type and concentration of the agent. The goal is to engineer small, inexpensive, easily deployable sensors to protect troops on a battlefield. The ultimate device would be a Dick Tracy–style wristwatch that could sense bioagents and perhaps even provide antidotes. The same sensors could defend the public. They could be deployed like smoke detectors in crowded, closed areas that would be enticing terrorist targets, such as subway cars, train stations, tunnels, indoor stadiums, and government office buildings. Sensors for chemical weapons such as mustard gas that burn people's lungs or skin (not biological ones that transmit killer diseases) are already deployed in several subway stations in Washington, D.C.

One experimental device unveiled in 2001 is a plastic, battery-powered unit the size of a pack of cigarettes, made by Cepheid in Sunnyvale, California. Airborne contaminants waft through a grating at one end into small interior channels filled with chemicals that amplify patches of different pathogenic DNA. A minute ultrasound generator cracks open the cells with intense sound waves, releasing their DNA. Small, thin-film heaters then heat and cool the DNA, in a commonly used genetic engineering process known as polymerase chain reaction, or PCR, which creates multiple copies of the DNA. The DNA then mixes with other reactive chemicals there, and the unit's light-emitting diode shines if a reaction occurs, alerting the person carrying it that a pathogen is present. The Naval Research Lab in Washington, D.C., is developing a similar sensor that uses magnetic fields to detect bioagents. Another device, from MIT's Lincoln Laboratories, contains immune cells that have been genetically engineered to interact with various pathogens, in a chemical reaction that will glow if intruders are found.

All these units face significant practical problems, however: how to keep the reactive chemicals from breaking down over time; how to fit many different chemicals that respond to the wide range of potential bioagents in a tiny space; how to keep fragile immune cells alive in hot or cold or dry conditions; and how to prevent false alarms that would be certain to create panic.

In addition to such small, distributed detectors, innovators are exploring various large-scale detection technologies. Late in 2001 the U.S. Postal Service began testing equipment that would be installed along automated mail-sorting lines.[22] The machines would zap passing envelopes with radiation or sterilizing gases, killing anthrax as well as other bacteria and viruses inside. At this point the postal service only irradiates mail destined for government agencies in Washington, D.C. Meanwhile, the service is considering plans to rapidly test dust from sorting machines for biological agents before mail is sent along its way. It also plans to install powerful vacuum systems to keep the machines from spewing out anthrax or other agents onto postal workers, who bore the brunt of the 2001 anthrax attacks. These fixes will cost hundreds of millions of dollars, and no one really knows how effective they might be. But the consequences of doing too little are far too great.

Some postal authorities argue that the anthrax mailings were only a tiny demonstration of the havoc that could be perpetrated. Only a few letters were mailed, and they were sealed tight, obviously intended to infect only the addressees who opened them. If letters were sealed loosely, and hundreds were mailed in different parts of the country, they could spew a trail of spores or other pathogens far and wide as they passed through the nation's sorting machinery. This threat is too serious to simply ponder. We must invest now in better early-detection systems.

Better detection technology would form one part of a larger, more coordinated national surveillance system—our crucial first line of defense—and the CDC is devoting many of its newfound resources to this goal. After the anthrax attacks, it became clear that the U.S.'s surveillance infrastructure was woefully inadequate, particularly in its ability to process human and environmental samples, coordinate lab tests, and communicate findings, advisories, and warnings to the general medical community. The $1.2 billion that Congress allocated to the National Institute of Allergy and Infectious Diseases was meant to fund basic and applied research into vaccines, early-detection devices, and other countermeasures. The CDC received $918 million, and was to funnel $865 million of that to states—nearly ten times as much as the agency had disbursed the year before.[23] The states were to upgrade their disease-surveillance systems, improve communication among doctors and hospitals, and beef up lab facilities so they could better detect potential bioterrorism pathogens. Part of the intention was that the changes would also strengthen the nation's ability to battle naturally transmitted diseases. Dividends already were reaped during the 2002 spread of West Nile virus. Mohammed Akhter, executive director of the American Public Health Association, told the press that states like Louisiana might not have detected the outbreaks as quickly as they did without the new measures in place. CDC leaders added that some of the West Nile outbreak's challenges were similar to those the CDC would face during a bioterrorism attack.

To help train clinical lab managers, the CDC's National Laboratory Training Network also sent out bioterrorism training and reference manuals to 4,600 clinical labs around the country, and by September 2002 it had held sixty-four courses at six field offices for hands-on instruction in rapidly identifying bioterrorism agents.

The other $53 million the CDC received was to expand the National Pharmaceutical Stockpile, a system of prepackaged medical supplies stored at various CDC facilities that sit waiting to be deployed during an outbreak. The CDC can deliver the supplies anywhere in the fifty states within twelve hours. The expansion would add more medical supplies, vaccines, and antibiotics useful in fighting bioterrorist attacks, as well as antidotes for chemical agents.

In addition, the CDC earmarked part of its regular budget for improving its communication with health officials, labs, and doctors nationwide by expanding certain information networks like the Health Alert Network and the Epidemic Information Exchange (Epi-X). The Health Alert Network supports state and local health departments to ensure that they have the equipment and expertise to participate in Internet communication. The Epi-X provides other agencies with up-to-the-minute information on outbreaks, and was used to spread word about the anthrax attacks and the West Nile epidemic. Part of the budget calls for a new Laboratory Response Network, in which at least one lab in each state could essentially provide the same testing and reporting capabilities that the CDC has for incoming samples from people and places thought to be hit by bioterrorism. Currently the CDC works with a loose coalition of independent private and public labs. The CDC can issue advisories and provide grants to help improve their capabilities but it can't institute changes or even set standards. The system was severely strained as it tried to handle the 125,000 human samples and 1 million environmental samples arriving at labs across the country during the anthrax attacks. Lab performance ranged from superior to abysmal. And none of it was fast.

The CDC itself had appeared slow and unsure of itself during the anthrax attacks. It wasn't set up to work efficiently with the FBI, CIA, EPA, or Federal Emergency Management Agency, all of whose efforts had to be coordinated. Even the CDC's culture needed to change. Instead of its usual modus operandi—waiting to inform the public until all the information about an outbreak is collected and analyzed—in a crisis the CDC would have to communicate on a daily basis to let the country know how things stood. It failed. During the anthrax scare, Anthony Fauci, the director of the National Institute of Allergy and Infectious Disease, emerged as the government's chief spokesperson, which was an embarrassment to the CDC. Never in its history had the CDC even held a press conference.

Julie Gerberding, the CDC's new director, has vowed to correct all that.[24] Gerberding is a physician who was acting director of the National Center of Infectious Diseases, a branch of the CDC. In a September 2002 press release she noted, "We have the people, we have the plans, and now we have the practice. We're building our knowledge and capacity every day to assure that CDC and our partners are ready to respond to any terrorist event." The CDC would also work with the new Department of Homeland Security, which will oversee a host of national agencies to coordinate the preparedness effort. Its Homeland Security Advanced Research Projects Agency will give grants to companies and universities working on detection and border-security technologies, while the CDC and NIH will continue assessing and funding health-related research.

Other initiatives may help as well, like the Working Group on Civilian Biodefense—the network of U.S. doctors, scientists, health officials, and experts from military, government, and emergency management agencies that is publishing an ongoing series of articles in the *Journal of the American Medical Association* on how to recognize and treat each of the Class A bioterrorism diseases.

The CDC's upgrade is encouraging. But its reallocation of so many resources to the bioterrorist challenge is controversial. Some experts worry that the shift has gone too far, that the agency, and the Department of Health and Human Services as a whole, is taking too much money away from the war against natural diseases. In the past few years the CDC has had an annual budget of about $6.6 billion to assess, control, and fight not just infectious diseases but chronic illnesses including heart disease, cancer, and stroke—the nation's biggest killers—as well as more-general national health threats such as obesity, and even occupational and environmental illnesses. It is a huge job of great importance. Forcing the agency suddenly to become heavily responsible for stopping bioterrorism, including, in effect, running the country's entire medical response to any outbreak, could tear its attention away from its fundamental purpose. Indeed, Congress's 2003 budget request for the CDC included a $57 million cut in chronic disease prevention and health-promotion programs and a $10 million cut in infectious disease control programs. Furthermore, being the lead bioterrorism responder will require the CDC to adopt a culture that must rush to judgment on thin data, issuing daily advisories based on what it happens to know at a given hour.

The issue of where the CDC's focus and operating mode should lie is complicated by the widely differing views about how clear and present a danger biological terrorism truly is. The U.S. anthrax letters killed five people and sickened seventeen, which is reprehensible, but hardly the country's worst nightmare; for example, fifty-five Americans die from the flu every day. Late in 2000 the Henry L. Stimson Center for public policy research published a widely quoted report criticizing the federal government for wasting $135 million to train national guard troops in how to respond to terrorist attacks, even though there was no hope they would ever arrive on the scene in time. Meanwhile, the report noted, the government had grossly underfunded programs with Russia to prevent 10,000 former Soviet biological and chemical weapons scientists from being recruited by terrorists and hostile states like Iraq and Iran, and had also underfinanced the most important means of detecting a biological attack: a nationwide disease surveillance system.

Still, scientists on the front lines of protecting the public would rather overprepare than be caught short in an emergency. One defender is Phyllis Della-Latta,[26] an associate professor of clinical pathology at Columbia University. In October 2001 she held a concurrent position as director of the clinical microbiology service at Columbia Presbyterian Medical Center in New York City—the huge lab that processes hundreds of cell cultures each day with bacteria and virus samples from sick patients throughout the extensive Columbia Presbyterian hospital system. Labs like hers form the vanguard that must rapidly identify all sorts of pathogens from any and all people who walk into the hospital system's front doors. In October her lab was overrun with the need to test patient samples for anthrax.

Della-Latta also sits on the special Bioterrorism Commission assembled by former New York City mayor Rudy Giuliani, and she has no patience for critics who underplay the threat of deadly pathogens. "I knew it was just a matter of time before inhalation anthrax showed up," she says. "Yet I'm even more worried about other agents, like smallpox, or an Ebola-like virus, or a combination of such things. We know genetic engineering has been going on in hostile states. When Saddam Hussein said, 'A pox on America,' he wasn't just spewing rhetoric, he was making a threat. And the terrorists' work is getting more sophisticated. It's getting harder to stay ahead."

The old measures that emergency-response officials claim will contain outbreaks like smallpox, such as putting early victims into quarantine, simply won't work, Della-Latta says. "Take a walk in the wild and crazy Bronx, or in densely packed Chinatown," she says. "There's no way you will quarantine people, much less an infectious disease they might be carrying." At a minimum, she says, America must stockpile vaccines against bioweapons that intelligence officers know terrorists are toying with. And ultimately, Della-Latta says, the real solution lies in destroying hostile dictators and terrorist organizations, which requires politics and militancy.

Almost everyone in the scientific and public-health communities agrees that tremendous resources should be spent to improve the early detection of biological warfare agents and the rapid testing of human and environmental samples. Their reasoning is simple: The same capabilities will greatly bolster our ability to identify and contain outbreaks of naturally occurring infectious diseases. An integrated system could alert us in days if a new wave of tuberculosis or yellow fever was spreading across the country, or if some awful new pathogen had arrived from abroad. Advisories about food contaminants—be it *E. coli* found in meat from a given processing plant or packaged food spiked with botulism toxin—or lab results indicating a new strain of influenza could immediately appear on the computer screens of doctors and health officials everywhere.

Indeed, any step that is taken to defend against bioterrorism should be designed so that it will also be an effective defense against natural pathogens. Otherwise, the terrorist threat will have the insidious effect of depleting our programs for detecting and combating naturally evolving diseases.

CHAPTER THREE

OUR AILING PATHOGEN ALERT SYSTEM: *Germs from Abroad*

Among the sinister threats we face in the pathogen war are deadly germs that infiltrate from abroad. They are arriving in more ways than ever. In record numbers, our family members and friends travel to faraway lands, where some of them consume contaminated water and food and are stung by infected bugs, then return carrying disease organisms. Visitors and immigrants arrive from other nations every day and may well be carrying disease organisms with them. Insects from hundreds or thousands of miles abroad stow away in the passenger and storage compartments of cars, boats, and planes, escaping into our surroundings once the vehicles reach their parking spots, piers, and gates back home. Plants and animals, healthy and sick, contaminate cargo on the same journeys, too. The diseases these myriad creatures carry to our doorsteps run the gamut from old scourges we thought we would never again see to brand-new, exotic pathogens we have never seen before—making it extremely difficult for doctors to figure out why a person who suddenly appears in an emergency room is deathly ill. These foreign agents present an especially challenging test of our public-health detection and prevention system.

The reemergence of old-world diseases in the West is striking. Malaria currently slays hundreds of thousands of people in Africa each year by clogging their brains' blood vessels, yet 1,000 to 2,000 people annually are diagnosed with it in the United States. Most of them become infected while abroad, but if local mosquitoes happen to bite them and then bite one of us, they could spread the pathogen. There hadn't been such a case in the New York City area for more than forty years—until 1991. That year malaria infected two residents in neighboring New Jersey who had not traveled outside their local area. In 1993, three New York City residents became infected. In 1999 two boys developed intermittent fever, chills, and sweating two weeks after attending Baiting Hollow Boy Scout Camp, at the edge of a marsh alongside nearby Long Island Sound.[1] A New York City artist developed the same malaria symptoms in 2002. In all these cases doctors used blood tests to figure out what was wrong, and prescribed malaria drugs, which are usually effective, to cure their patients. But they were alarmed at what they were fighting. Such cases are likely to occur again, more frequently and widely, because malaria is resurgent overseas; from 1998 to 2000, malaria killed as many people globally as AIDS.

Cholera is also returning, poisoning shellfish along U.S. shorelines and endangering anyone who eats them. The bacteria causes profuse watery diarrhea and vomiting, and even death, and can be spread through contaminated water or raw or uncooked shellfish. Freighters from abroad may be the couriers; to clear the seafloor in shallow port waters, big boats may release up to 79 million tons of ballast water, taken in on distant shores, that might harbor the pathogen.[2] Disease-causing cholera organisms have been found in the coastal waters of Maryland as well as in shellfish harvested from Mobile Bay on Alabama's Gulf Coast.

More troubling still are newly evolved diseases that have made their way here. Unfamiliar to our public-health defenders, these exotic pathogens unveil the weaknesses in our piecemeal system for detecting and responding to infectious diseases. No case makes this clearer than the 1999 West Nile outbreak in New York City. This remarkable story shows just how much luck was involved in successfully identifying the disease quickly enough to minimize its deadly effects. And yet, even with that luck, the disease is spreading rapidly. After simmering for two years, the virus infected more than 3,600 Americans in 44 states in 2002 and killed 212. West

Nile virus is also a chilling example of how an exotic foreign disease might arrive on our doorstep any day.

Enrico Gabrielli arrived home hot and tired on the sweltering Wednesday evening of August 11, 1999.[3] His job at a mannequin factory in Elizabeth, New Jersey, was sweaty enough, but his long subway ride home to the New York City borough of Queens was almost unbearable. The sixty-year-old Italian immigrant got off the train at the Whitestone stop and made his way to his equally hot little brick house. He was glad to be home. His wife, Caterina, had cooked dinner, and the couple ate outside at the old table in their tiny backyard, as they frequently did in the summer. It was cooler there, and they could survey the adjoining city backyards, filled with tomato plants and hanging laundry.

As daylight faded after dinner, Enrico would usually water the vegetable garden or putter around, fixing this or painting that. The only drawback was an occasional mosquito bite, but that was a small price to pay for the pleasure of being outside. On this evening, though, Enrico wasn't feeling well. His head hurt. He was having chills and fever. He swallowed two aspirin tablets and went to bed. His fever got much worse overnight, spiking to nearly 104 degrees several times, interlaced with severe chills. His muscles quivered. By the next morning his limbs, already aching, began to become rigid; he found it hard to walk. He called in sick to work, then his legs collapsed under him in the kitchen. Caterina immediately phoned for help, and an ambulance rushed him to Flushing Hospital Medical Center. By the time they wheeled him into the emergency room, Enrico could not move his arms or legs at all. It was as if he were suddenly paralyzed. He also seemed disoriented and confused.

The ER physicians infused Enrico with strong drugs to try to relax his stiffening body, but they did not know what was attacking him. High fever suggested infection, but Enrico's near-paralyzed state indicated a completely different set of possible illnesses. As his muscles tightened, the doctors grew more tense; if his chest seized up, he would stop breathing. They moved Enrico to Intensive Care, where Caterina stood anxiously by his side. For a week the IC staff fed intravenous fluids into Enrico's veins, which helped to lower his fever, and gave him drugs to try to relax the nerves that were contracting his limbs, but the symptoms didn't lessen. So they called in Deborah Asnis, a physician and infectious disease specialist; perhaps Enrico had some strange illness no one had foreseen.

By the following Monday, August 23, Asnis had become worried that Enrico might in fact have some exotic disease. Another patient, a seventy-five-year-old man also from Whitestone, had arrived in a disoriented state, unable to move his limbs. Asnis had ordered a spinal tap for each man, and the lab results indicated they both had high levels of white blood cells in their spinal fluid—a sign that the nervous system was under siege. Asnis suspected viral encephalitis, a dangerous inflammation of the brain, or Guillain-Barré syndrome, which has similar symptoms. She gave the patients acyclovir and cephalosporin, standard treatments for these possibilities, but the patients worsened. The symptoms didn't make sense, either; encephalitis doesn't cause muscle weakness, and Guillain-Barré doesn't usually create a high white-blood-cell count. She thought that perhaps botulism was at fault. It produces toxins that can prevent nerves from communicating, and can cause victims to lose muscle control, although it is not usually associated with a change in mental state.

If the two patients did have botulism, that meant that there might be lots of ticking time bombs out there—store shelves full of toxic food. Many other people could be exposed. So on that Monday, Asnis did something other doctors might not have bothered to do. She called Marcelle Layton, assistant commissioner of the New York City Health Department, located in a formidable marble building in lower Manhattan, across from City Hall. Layton, a physician trained in infectious diseases and hospital epidemiology at Yale, oversees the health department's Bureau of Communicable Diseases.[4] Because the city is so huge and is also the entryway for many immigrants, the health agency has an entire bureau dedicated to each of three globally widespread ailments: HIV, tuberculosis, and sexually transmitted diseases. Every other infectious disease that arises within the city—fifty-three reportable ones in 2002—falls under one burdened bureau: Layton's.

According to state health-code law, doctors in New York are supposed to report to their local or state health department any time they find one of dozens of dangerous diseases—from anthrax and botulism to malaria and yellow fever. The same code also states that, more generally, a physician should report any unusual manifestation or cluster of illnesses. But one of the worst problems with our disease-detection system is that many doctors never report cases of strange symptoms, either because they are unsure of the disease they are facing, they're ignorant of the reporting requirement, or they simply never get around to it. This lone lapse in a doctor's duty can let loose an outbreak. Fortunately for many New Yorkers, Deborah Asnis was highly conscientious.

Asnis told Layton about the unusual patients. The two women reviewed the symptoms and lab results, and agreed the data didn't add up. They were stumped. They agreed that Asnis would send blood and spinal fluid samples to the New York State Health Department labs in Albany. They were lucky they had this option. Hospital labs test for the most common of organisms, and city labs tend to test for less common pathogens like the one that causes botulism. State labs test for diseases that are more unusual still, with the CDC attacking the truly mysterious germs. Few state labs were equipped to test for arboviruses—those carried by ticks or mosquitoes—but New York happened to be one of only three state labs in the country that had received CDC funding to set up the sophisticated arbovirus tests.

By Friday, August 27, Enrico's condition was no better and more patients with the same symptoms had arrived at the hospital. Asnis became alarmed. An eighty-year-old man and an eighty-seven-year-old woman from Enrico's neighborhood had been admitted. The man had had a heart attack and could not move his muscles. The woman had become essentially paralyzed. A neurologist whom Asnis knew told her he was treating yet another patient with similar problems in a nearby hospital.

Asnis called Layton again. Even if the five patients' parallel symptoms weren't encephalitis or Guillain-Barré or botulism, they certainly fit the criteria of an unusual cluster of illnesses. Both women were leaning more toward encephalitis; the city typically saw nine or ten such cases a year, and to have five from the same small section of Queens at the same time was striking. None of the patients was doing well, either. Two were on respirators. Enrico was breathing on his own, but his body was rigid. He slept motionless most of the day, and was delirious during the brief periods he was awake. Caterina worried Enrico might be paralyzed for life, that he would live out his days as a vegetable in a hospital bed.

Layton was not about to allow herself to be stymied by the cases. She tackled her job with endless energy, and on that Sunday she drove over to Flushing Hospital with her colleague, Annie Fine, to study the patient charts herself. She was not a practicing physician but had the training; she wanted to examine the cases firsthand and talk to the patients and their visiting families. While she was there, a fifty-seven-year-old man from a neighborhood near Whitestone was admitted, hallucinating fiercely. His condition frightened Layton; that kind of delirium strongly signaled encephalitis. She ordered her staff to check records from other Queens hospitals, and they turned up two more patients in equally dire straits. Layton then ordered more samples from those patients to be sent to the state labs.

Eight victims with the same severe symptoms, all from the same few square miles of Queens? Layton didn't know what was going on in Whitestone, but she wasn't about to wait any longer to find out. The next morning she called the Centers for Disease Control headquarters in Atlanta. She needed help.

The CDC is often called "the nation's doctor." It is a major part of the federal Department of Health and Human Services, and comprises various centers, including the National Center for Infectious Diseases. The CDC provides expertise, funding, and occasionally personnel to state and city health departments as special needs arise, but the local authorities have to request the CDC's help before it can become involved. Layton was not reluctant to ask the CDC's experts to make a house call. And she knew who she wanted: someone from the Epidemic Intelligence Service.

In 1951, following the Korean War, CDC cofounder Alexander Langmuir formed a corps of young medical people, dubbed the Epidemic Intelligence Service, to fan out across the country to provide early warning if biological warfare attacks were made.[5] Soon the CDC was dispatching EIS "officers" around the United States and the world to examine weird outbreaks and fight disease on its front lines. It was EIS officers who actually discovered that a bad batch of vaccine from Cutter Laboratories, made to fight a frightful American polio outbreak in 1955, was actually causing paralytic polio in fifty-nine vaccinated children and 133 unvaccinated family and community members; an additional sixty-eight people ended up with a form of the disease. They discovered that lead paint was poisoning children in 1971; that airborne bacteria was causing Legionnaire's disease in 1976; that tampons were causing toxic shock in 1980; and that *E. coli* O157 was what sickened hundreds of Jack-in-the-Box patrons in 1993. EIS officers even helped discover both the HIV virus and the Ebola virus.

Today, EIS officers still watch for bioterrorism and infectious diseases, but their role has expanded to investigate all types of epidemics, including chronic diseases, obesity, and injuries. Each year the CDC selects a class of about sixty-five doctors, researchers, and scientists to fill two-year EIS terms. Most are young physicians eager to sleuth medical emergencies. Sent to the scenes of suspected outbreaks, these "disease detectives" live out of suitcases, always on call, at times putting their own lives at risk in the face of unknown pathogens. The CDC's chief in 2002, Jeffrey Koplan, cut his teeth as an EIS officer. So did Marci Layton—yet another stroke of luck in this confounding case. Layton's EIS experience made her very aware of the need to inspect minute details of an unfolding outbreak to determine what was really happening—and more ready to quickly call in the CDC instead of relying just on her own, overworked staff.

It so happened that an EIS officer was already on loan to New York City for other duties. But the man was on vacation. That's when Denis Nash's phone rang. The inexperienced, thirty-one-year-old EIS recruit had just been assigned to New York also, to help the health department track new cases of AIDS. Two months earlier Nash had graduated from the University of Maryland with his medical doctorate in chronic diseases, and hadn't even moved his personal belongings into his newly rented Brooklyn apartment when he received the call from the EIS manager in Atlanta whom Layton had reached. The manager told him about the cluster of illnesses. He gave him the addresses of the eight patients, and told him to ask Layton for several specialists to help him. He also told Nash to follow the EIS's standard procedure: Comb every square foot outside each person's residence and every square inch inside for anything that might provide a clue to disease. Then interview the families.

Neither the manager nor Nash, however, really had any idea what evidence could be the smoking gun.

On Wednesday morning, September 1, Nash arrived at Enrico's house in a health department van and parked in front. The home had that classic Queens look made famous by Archie Bunker's residence in the TV series *All in the Family.* Inside the van were four people: Nash, an exterminator, an animal-disease expert, and an entomologist from the American Museum of Natural History named Varuni Kulasekera. Each of them had a trained eye for spotting unusual activity in bugs or vermin or even pets, as well as other clues, such as spoiled food or unsanitary conditions. They fanned out around the house, looking at, under, and into every nook and cranny. Then they went inside.

The crew repeated the same exercise at the second patient's house, then the third, and so on, all day. They talked to whichever family members were around. Back at Flushing Hospital, Asnis hung her head. The eighty-year-old man who had had a heart attack and had been on a respirator had died that same afternoon. She feared that the old woman, who had been admitted after Enrico, might succumb as well because her brain was now extremely inflamed. Enrico's limbs remained paralyzed, but at least he could breathe on his own, and the swelling inside his head was not as extreme.

The next day, Thursday, evidence finally began to come together. And it was none too soon. Just as Asnis had feared, the old woman died that morning. Asnis also learned that more than thirty people with similar symptoms had been admitted to area hospitals. She now worried that a major outbreak was in the works. Kulasekera had noticed standing water at several of the patients' homes. At the house of the eighty-year-old man who had died, she had found five-gallon buckets in the backyard, which the man had used to collect rainwater for his garden. The man's wife and son said he had trouble sleeping, and usually got up at dawn to smoke outside. He smoked after dinner, too, as the sun went down, often standing beside those very buckets. The EIS team found rain buckets outside another house as well. They knew that standing water was a prime breeding ground for mosquito larvae, and Kulasekera had found *Culex pipiens* larvae floating there. *Culex pipiens* is a mosquito species that bites most aggressively at dawn and dusk, and Kulasekera knew it could carry encephalitis viruses. That same afternoon the state lab called Marci Layton: the blood and spinal fluid from Enrico and other patients had tested positive for St. Louis encephalitis. The mystery appeared to be solved.

St. Louis encephalitis is caused by a virus that was first noted in an outbreak around St. Louis in 1933. It has since spread throughout the United States. It belongs to a small group of viruses that voyage back and forth among birds and mosquitoes. Man is an accidental target. The virus multiplies inside bird cells, but depends on mosquitoes to transport it from bird to bird. Occasionally, when a mosquito bit an infected bird, it would pick up the virus, and the virus could dwell in the bug's saliva. Then, when the mosquito jabbed its beak into a human for its next blood meal, it could inject the saliva into the victim's bloodstream.

Layton did not rush to issue a public alert, however. She knew from experience that the test the state lab had used was a simple one, and not terribly accurate. It was useful in a preliminary way to help focus in on a pathogen, but wasn't reliable enough for a definitive diagnosis. She also knew that there were different forms of encephalitis that could be mistaken for one another. For precisely this reason, she had also sent blood and spinal fluid samples to a CDC lab in Fort Collins, Colorado, which specializes in mosquito-borne diseases. The results were due in a day, so she decided to wait to hear from the CDC as well before going public with the news. The next morning Duane Gubler, a director at Fort Collins, called Layton and told her that his lab's initial tests had also come up positive for St. Louis encephalitis. He was concerned about that finding, though. CDC scientists had never seen a strain quite like this one before. He was going to run more-specific genetic tests.

Layton was now growing concerned that a St. Louis encephalitis epidemic was under way. Although no outbreak had ever occurred in the New York area, maybe mosquitoes or infected birds had brought the disease there. The CDC advised her not to wait until the results of the additional tests came back to notify the public. Layton agreed, but also decided to order yet more-detailed analysis from the New York State lab; the more viewpoints she received the better.

Layton felt uncomfortable putting the city on alert that an outbreak was under way without knowing for certain it was St. Louis encephalitis. But she had to say something—precautions had to be taken. So on September 3, New York City mayor Rudolph Giuliani and one of Marci Layton's deputies held an afternoon press conference at City Hall. They said a half-dozen patients were deathly ill in several city hospitals. State and CDC lab tests had shown that an outbreak of St. Louis encephalitis had begun in Queens and could potentially crisscross the city, because it seemed that mosquitoes were transmitting the infection to people. The mayor reassured potentially frightened citizens that the city's health department was on the case and had set up a hotline for people with similar symptoms. He also announced that the city would begin spraying certain neighborhoods in Queens with insecticide, to cut down the mosquito population. Helicopters would be spraying Whitestone from above before dinnertime.

At the end of the press conference a woman approached Layton's colleague, Annie Fine, and asked, "Is this related to the crows dying?" The question surprised Fine, and Layton as well when Fine later relayed it to her. Layton's office had been unaware that since June, residents from Queens and the Bronx had been calling various wildlife agencies to report that an unusual number of dead crows were turning up on city streets. As was all too typical in cities across the country, those agencies hadn't informed the health department, and the health department didn't monitor those agencies' general reports, either. Back at her office, Layton checked with the CDC and was told that St. Louis encephalitis didn't usually kill birds. So she assumed the crow deaths were coincidental.

The only local government employee who had perceived that something was terribly wrong with the birds in the area didn't work for any of the health agencies. She was the head pathologist at the Bronx Zoo. Tracey McNamara was responsible for the welfare of all of the zoo's animals.[6] By early August she had found more than forty dead crows around the zoo. It was not uncommon to see die-offs during migration season, from the stress of the arduous flights, but it was nowhere near that time of year. Worried the crows might be dying of a disease that could spread to the other zoo birds, McNamara had sent several dead crows to the State Department of Environmental Conservation, asking that lab tests of their tissue be done. She had received back only a general autopsy report, which told her nothing.

When McNamara heard about the city's press conference, she began to wonder about the hypothesis of encephalitis, and a link to the dead birds. Then, over Labor Day weekend, a Chilean flamingo, a Guanay cormorant, a pheasant, and a bald eagle at the zoo all collapsed and died. McNamara was heartbroken—and angry at the mysterious foe. Deciding to do her own biological sleuthing, she put on her surgical gown and face mask and autopsied the dead birds herself. What she discovered stunned her. She found hemorrhages in their brains and striations in their hearts. Slicing off paper-thin sections of the diseased organs, she placed them on microscope slides and dyed them. Looking through the microscope at the stained tissue wafers, she saw the worst encephalitis damage she had ever seen.

McNamara was sure there was a link between the avian and human outbreaks. But as Layton had been told, even though birds can carry St. Louis encephalitis, they don't die from it. Now McNamara was driven; more of her precious birds were falling ill, and a link between the city's crows and human victims could spell widespread trouble. McNamara felt that city officials were wrong about the St. Louis virus identification, but she needed data to back up her suspicions before confronting them. So she carried on an independent investigation. On September 9 she called the CDC in Fort Collins to see if they would look at her bird samples, too. They said no; birds were not their business. Frustrated, she called the National Veterinary Service Laboratory in Ames, Iowa. Technicians there said they would test her specimens. She also called a colleague at USAMRIID, the military disease center. She wanted to cash in a favor; he agreed to test her samples, too.

The lab work being done for Layton and McNamara was painstaking and therefore slow. Asnis and Layton went crazy waiting. All of the first four victims had died except Enrico, who was actually beginning to do a bit better. Doctors had recently moved him out of Intensive Care and into a regular room. He was still on a catheter, still had occasional hallucinations, still could not walk on his own, and had lost twenty pounds, but he was awake more, and could control some of his muscles. Caterina was hopeful.

More and more city residents, however, were turning up at hospitals with worrisome symptoms. By the third week of September, more than forty people had been diagnosed with encephalitis. Layton was feeling uneasy as well as impatient. Results began arriving from the advanced tests at Fort Collins, and they only weakly indicated St. Louis encephalitis. The city's tabloid newspapers had connected the human and avian stories, even if the experts had not declared there was a formal connection. People from across the five boroughs had started calling Layton's office reporting dead birds. And CDC personnel had begun coordinating communication among Layton, McNamara, and all the labs involved.

Finally the national veterinary lab reported to McNamara and the CDC that it had found a virus in McNamara's birds. They identified it as being in the same family of viruses as the St. Louis virus, but they didn't have the capacity to pin down exactly which virus it was. McNamara's friend at the army lab also called her and said he had found evidence that the virus was in the same family of pathogens. The CDC had been unable to grow the virus from human samples because there was simply too little virus there. So now they became extremely interested in the bird samples. Robert Lanciotti, a colleague of Gubler's at the Fort Collins lab who specialized in immunology and genetics, asked the veterinary labs for a sample. He used a highly sophisticated test to investigate the proteins that immune cells in the deceased specimens had created, and was able to rule out St. Louis encephalitis virus. But he still couldn't define exactly what the pathogen was, so he decided to sequence parts of its DNA.

Once Lanciotti finished the sequences, he e-mailed them to a database of genetic sequences maintained by the National Library of Medicine, to see if they matched other gene maps stored there. It was Thursday, September 23, five weeks after Enrico had fallen to his kitchen floor. The result came back to Lanciotti's computer quickly. His sequence matched that for another pathogen already in the database: West Nile virus. He was shocked.

Microbiologists first isolated this peculiar virus from a woman in the West Nile district of Uganda in 1937.[7] It is genetically close in form to several other mosquito-borne viruses found around the world, including St. Louis encephalitis, Japanese encephalitis, yellow fever, and dengue fever. The West Nile variety caused outbreaks of encephalitis in Israel in the 1950s, in South Africa in 1974, and in Romania in 1996. Each time hundreds of people became ill, and 5 to 10 percent of them died. West Nile, like the related viruses, was not spread from person to person, but by mosquitoes that bit infected birds and then bit people. But it had never appeared in North America. Amazingly, the CDC had tests for the virus that it had developed while helping the Romanian authorities with their outbreak, tests that could have identified the virus from the Whitestone victims within days. But the CDC had not considered West Nile virus a possibility.

Could West Nile virus really have landed in New York City? Layton, McNamara, Asnis—everyone—was stunned. But confirmation came immediately. A week earlier, in yet another twist of fortune, one of the state lab scientists to whom Layton had sent samples was at a conference where he met a specialist who was renowned for his genetic diagnoses of diseases—a molecular biologist at the University of California, Irvine, named Ian Lipkin. The two men hit it off, and the lab scientist sent Lipkin samples from the two patients who had died. Lipkin rapidly sequenced a portion of their genomes.[8] He used the Internet to check databases for genetic matches against five viruses that he knew had been sequenced and caused encephalitis. One match came back, and it indicated that the pathogen was Kunjin virus, a West Nile–like virus found in Australia. On that same Thursday he called the state lab and the CDC officials. In the following days when Lipkin looked at larger genetic stretches of the virus, he, too, agreed it was not merely *like* a West Nile virus, but *was* West Nile itself.

Two independent investigations had reached the same conclusion. Hard as it was to believe, New York City was under attack by a virus that had arrived from an ocean away.

Marci Layton's astonishment quickly gave way to concern. West Nile virus reproduced widely in birds, so she expected that a great many would be infected. She also knew that *Culex pipiens* mosquitoes, the kind the EIS officers had found, were good at carrying encephalitis viruses like West Nile. Unfortunately for all, *Culex pipiens* was the common house mosquito, the most prevalent species buzzing around New Yorkers. September was also peak mosquito season. Even though it might be uncommon for a mosquito to bite a bird and then a person, Layton knew it could happen frequently enough to spread the disease quickly.

In a model public-health campaign that other cities should learn from, Layton's office rapidly organized a publicity program to inform city residents that rather than St. Louis encephalitis, the disease that was infecting people was the obscure West Nile virus. The advice stayed the same, however: They must avoid mosquitoes. They should avoid the outdoors at dawn and dusk, wear long-sleeved shirts and long pants, and clear standing water from their houses and yards, even seemingly small, innocent puddles in flowerpots and birdbaths. Layton also arranged for the city to give out insect repellent free.

Mayor Giuliani ordered city workers to step up the spraying—the most massive bug-spraying campaign ever waged in an American city. They were to blanket Queens with the pesticide malathion, which kills bugs by crippling their central nervous systems. Every day, people shut their windows and ducked inside their homes when they heard the oncoming spray trucks rumble up the street. Helicopters sprayed malathion from overhead onto open areas. Residents called the health department not to thank them but to complain of headaches, nausea, vomiting, blurred vision, and fatigue. A debate flared up on the streets and in the news about which was worse, West Nile virus or the poison being sprayed to stop it.

With the end of mosquito season in late November—after several hard frosts—public-health officials knew the spread of the disease was over for that year. They also knew the virus had probably infected many more birds than it had killed, and that it might well flare up again come next mosquito season. Seven New Yorkers had died in the outbreak.[9] More than fifty others, most of them senior citizens, had been sickened with ailments ranging from intense headache to meningitis—inflammation of the lining of the spinal cord and brain. Enrico was one of the worst afflicted of the survivors. His central nervous system suffered severe damage. Three months after being bitten randomly by a seemingly harmless mosquito, he still could not walk unaided, dress himself, or raise his right arm over his head, despite ongoing physical therapy.

Enrico was not the only one to suffer long-term complications. In December 1999, four months after Enrico's case of West Nile appeared, a study done by Layton and the CDC showed that 70 percent of West Nile patients who had survived still had muscle weakness, 75 percent had memory loss, and 60 percent suffered from confusion.[10] Half the victims were age sixty-eight or older. Autopsies of those who perished showed they died primarily from encephalitis. But, oddly, not a lot of virus was present in their brains, nor was there much visible brain damage, as McNamara had found in the dead birds. It seemed that in humans, most of the infection and immune war took place in the brain stem, at the base of the skull behind a person's ears. This area controls such basic, involuntary functions as breathing. As the immune war is waged inside each human, even minor damage to the brain stem could cause the victim to stop breathing or fall into a deadly seizure or coma.

By the time the summer of 2000 arrived, some victims, like sixty-six-year-old Bob Benson of Bronxville, had fully recovered, although they were uniformly afraid to go outside without drenching themselves in bug spray. Others were not so lucky. Danuta Trojanowska, fifty-six, of Port Chester, still could not walk the length of New York United Hospital's fifth-floor hallway, because of significant neurological damage. Enrico still had trouble using his right arm and left leg, and was still too weak to return to work.

Marci Layton worried about whether the virus would return. This time the city was set up to do its own testing. A springtime survey of birds near the epicenter had revealed that up to 67 percent of crows and 60 percent of house sparrows were infected in some neighborhoods.

Layton soon received more evidence that the virus would come back. Over the winter, one of her deputies, James Miller, in charge of mosquito-borne diseases, and several EIS officers had found the virus in mosquitoes that were thriving in warm havens like the city's subway system.[11] By June the killer mosquitoes were laying their eggs in water barrels, rain gutters, storm drains, pool covers, discarded tires—wherever water stood. Giuliani dispatched a brigade of fifty-four city workers and seven supervisors to spread larvicide in the city's 135,000 storm drains, switching the personnel from their usual job of cleaning vacant lots to control rats. The city hired a large squad of college students to find and eliminate breeding places in ponds, streams, and man-made containers. The larvae are about one-third of an inch long and can be readily spotted with the human eye that knows what to look for—a huge saving grace. Were the larvae much smaller, finding them would be extremely difficult.

Layton's health department, and those in adjacent Long Island to the east, Westchester County to the north, and New Jersey to the west, also deployed dozens of chickens as sentinels—confined in coops as live bait for biting mosquitoes. Chickens do not get ill from the disease because they produce a strong antibody response. Researchers drew their blood every two weeks to see if the antibodies were present, meaning that the fowl had been bitten by infected mosquitoes. Under direction from the CDC and the state, the city also hung hundreds of mosquito traps—boxes baited with a disgusting sort of emulsion—in trees and on lampposts. Health department workers collected the dead bugs and sent them to the state lab for testing. They collected more dead crows and sent them, too. Layton's office also launched a massive public-awareness campaign; her employees spread flyers everywhere and ran print, radio, and TV ads once again explaining how people could limit their risk of being bitten.

Despite all these measures, the disease quickly spread beyond New York City. By November 2000, dead crows, blue jays, and sparrows carrying West Nile were found in parks, along roads, and on people's lawns from Boston to Maryland. Horses in several states were infected. Sick bats, a raccoon, and a chipmunk tested positive. In addition to *Culex pipiens,* seven other mosquito species were now documented as carrying the virus, including *Aedes alboictus*—the Asian tiger mosquito—an aggressive biter also capable of transmitting yellow fever. New Jersey declared its first human victim of West Nile. The autumn migration of fifty-three species of infected free-ranging birds from eight Northeastern states would certainly move the disease into the Southeast.

As with most infectious diseases, people differ markedly in how well their immune systems respond to West Nile virus. After the initial outbreak in 1999, Layton and the CDC's EIS officer Denis Nash surveyed residents and took blood samples from the general population at the epicenter in Queens and several other city areas.[12] They estimated that 8,200 people had been infected. From interviews they found that about 20 percent had experienced fever, headaches, muscle and joint aches, and fatigue between late summer and early fall. Less than 1 percent were hospitalized. The most important factor in contracting the disease was age. One in fifty people over age sixty-five developed serious illness, compared with one in 300 of those under sixty-five. Diabetes also increased susceptibility. Luckily, children, perhaps the group most likely to spend summer days outdoors, seemed no more likely to become ill. Only three of the patients were under the age of sixteen.

As West Nile virus spread farther across the United States, it was hard to imagine it disappearing in New York City. Yet in 2001 it was relatively quiet there. The aggressive campaign against mosquitoes had paid off. It continued to spread, however. In August a woman who lived near downtown Atlanta, not far from CDC headquarters, died from it—the first reported death outside the Northeast. Dead crows were found in Ontario, Canada, and a dead blue jay was found in Cleveland, Ohio—a disturbing sign. Blue jays generally don't migrate; they stay close to home, meaning that a bird or an insect must have brought the disease to the Ohio region.

The disturbing extent of the disease's migration was made dramatically more clear in the summer of 2002. Mosquito-borne diseases tend to be episodic—they flare up and die down—and in 2002, West Nile went wild. By the end of December there had been more than 4,100 cases and 270 deaths in forty-four states and the District of Columbia, with Illinois and Michigan the hardest hit.

Three years after West Nile virus reached the United States, microbiologists have learned a great deal about it, but are still not sure how it kills people. Somehow it triggers the brain swelling that shuts down the central nervous system, causing victims to die. There is no vaccine or cure. We have simply been lucky that our immune system seems to be able to fight it—unless we are elderly.

Epidemiologists, for their part, still don't know how West Nile virus made it to America. Since Colonial times, New York City has been attacked by diseases from abroad. In the 1660s, officials running the nascent port city kept British ships in quarantine to prevent plague, then sweeping through London, from coming ashore. Smallpox arrived on slave ships in 1689. Other ships brought measles, scarlet fever, typhoid, and malaria. A yellow fever epidemic in 1743 wiped out 5 percent of the city's residents. Malaria and yellow fever, transmitted by mosquitoes, had also struck the city. Theories about how West Nile arrived abound: Some say sick goats were shipped from the Middle East through Turkey to the United States. But scientists have no conclusive evidence at all.

The virus might well have come to New York in ships or planes from abroad, but another thought also occurred to government officials in 1999: Could the virus have been released by a terrorist? There had been reports that Saddam Hussein had threatened to use West Nile virus as a weapon in 1997. It was Layton's responsibility to find out. Any time an unusual disease appears in an unusual cluster, Layton's office must suspect bioterrorism and research and prepare accordingly. In fact, Layton had already met with FBI and CIA agents by the time Mayor Giuliani announced the West Nile outbreak to the public. But no credible intelligence information came forth, and no group took credit for an attack. And even though West Nile was on the CDC's list of potential bioterrorism agents, many other diseases were far more deadly, and thus better choices for a would-be terrorist.

After a considerable investigation, the CDC's best guess was that the New York outbreak might have come from Israel. The American strain was virtually identical to a virulent strain that had caused a recent outbreak in geese there. A mosquito could simply have drifted into an airplane before its door was closed at Tel Aviv's airport and escaped again once it was down in New York eight hours later.

The notion that mosquitoes can spread infectious diseases by hitchhiking on planes is all too real. In August 2000 the World Health Organization issued a warning about "airport malaria" spread by stowaway bugs. In 1994 six French residents living near Charles de Gaulle Airport in Paris came down with malaria, even though they had never visited a country where the disease prevailed.[13] Late in 2000, Martin Cetron, an epidemiologist at the CDC's quarantine division in Atlanta, determined that from July 1998 to June 1999, 5 million people from overseas flew into the New York metropolitan area's airports, and of them, 2.1 million had come from places where West Nile had been found. Horses, frogs, and more than one million live birds and chickens—all of which can carry the virus—had arrived in that time span, too. Such are the risks of a shrinking globe.

Recent experiments also indicate that we may not yet fully understand how the West Nile virus is spread. Researchers at the National Wildlife Center made a startling discovery that it could even be contagious, when it spread in a lab that was free of mosquitoes.[14] The scientists deliberately injected crows with West Nile virus, and all of them died. Unexpectedly, however, nearby crows that had not been injected also became infected and succumbed.

The researchers theorize that the virus may have spread via saliva when the birds cawed or pecked at one another, or stole one another's food.

There is no suggestion that under normal circumstances people can catch West Nile virus from another person. But in late August 2002, CDC experts discovered that four people might have contracted the virus when they received contaminated organs donated by a Georgia woman who had died in a car crash. One recipient, an Atlanta man, later died. The CDC also found evidence suggesting that a half-dozen people in four Southern and Midwestern states might have contracted West Nile from transfusions of blood donated by infected people.[15] Immediately in New York City, there was a 50 percent rise in the number of people expecting to need transfusion who sought to pre-donate their own blood. In September the CDC was scrambling to pin down these presumptions while it also developed a test to screen the blood supply. In November 2002 a woman who had been infected with West Nile virus while she was pregnant gave birth to an infected infant, who showed severe brain defects. There was also an infant who contracted the disease from her mother while breastfeeding.

The story of the onset and rapid dispersion of West Nile virus in the United States chillingly illustrates our vulnerability to emerging diseases from abroad, and points out the many weaknesses in our current system for detecting such foreign agents. Deaths were limited in New York City's West Nile outbreak in part because, through sheer luck, capable, persistent people were in just the right places at the right time.

It was good luck, for example, that Asnis called Layton to report what, at the time, were only two odd cases. Asnis was much more conscientious than many doctors. Layton also knew, through personal experience, how much the EIS could help, and she actually sought the CDC's aid, instead of failing to call out of some foolish pride that her department didn't need any meddlesome outsiders. It was even more fortunate that McNamara took the extra step to doggedly autopsy so many dead birds under her own microscope. She wasn't deterred when a close-minded Fort Collins lab turned down her request to analyze those samples for encephalitis, and she was lucky to have a personal friend at USAMRIID who did the tests instead. And it was plain dumb luck that disease specialist Ian Lipkin was attending a conference that happened to be under way in the third week of the crisis and not at any other time of year.

Finally, it was luck that the city's timely response to the erroneous diagnosis of St. Louis encephalitis—spraying for mosquitoes—happened to be the right public-health defense for West Nile virus, too. The health department's decision to announce so early the increasing cases of encephalitis was also crucial. As Layton herself now points out, "I always wonder, if we hadn't gone public early, would people have looked so closely at the birds?" It's those politically tricky decisions that can mean the difference between an undiagnosed tragedy and a quickly resolved health situation.

Clearly the United States cannot rely on such an almost unbelievable string of good fortune to defend itself against the disease threat. If the spraying hadn't begun so early, thousands more mosquitoes would have fed on hundreds more dead crows and transferred the unknown virus to scores more people—at a minimum. And what if the outbreak had begun in another big city, like Baltimore or Charleston or Tampa, that wasn't located in a state that had one of only three advanced labs in the country that could test for such diseases at the time? By now most, if not all, state health departments can test for West Nile virus.

So what have we learned? Surveillance is one part of the answer, but it can be hard to put effective surveillance systems in place. To monitor for West Nile, there are now enclosed pens holding from two to thirty "sentinel chickens" up and down the East Coast and beyond, and health departments have put up mosquito traps in numerous places, all to create an early-warning network for returning or spreading virus. This system requires personnel to check the traps, and lab facilities to test what they collect. All of which requires money.

The West Nile monitoring system was established, of course, only because of all the public furor in 1999. Layton's health department received city and federal funds to build the infrastructure. Because West Nile is seasonal, during other times of year Layton assigns the same staff to monitor Lyme disease, which requires the same lab capabilities. "I wish we could get the same infrastructure for other diseases," Layton says. "I'd like to get it for hepatitis A and B and C; the burden of these diseases out there is great, but we're getting no additional funding from the city, state, or CDC. A solid infrastructure requires trained people who can go out and review cases, interview patients, review medical charts, and oversee a database to manage all the data."

Bad as the epidemic was in the 2002 season, it would have been worse if an early-surveillance system hadn't been in place to beef up mosquito-control measures at the first sign of spread to a new locale. Indeed, the West Nile epidemic demonstrated glaring shortcomings in the preparedness of the public-health system to meet emerging infectious disease challenges. From the time the public-health system was founded—in New York City in the 1860s—public and political support has waxed and waned as new diseases rose and subsided. In the 1960s and 1970s, for example, New York had a strong monitoring program for mosquito-borne disease, but officials allowed it nearly to collapse, and it turned its attention to more pressing—or politically fashionable—threats. In 1998 the mosquito-borne-disease program for the entire state was staffed by the equivalent of one and a half full-time people. Across the country there was a shortage of people trained in arboviruses, because that discipline seemed old-fashioned compared with the specialties confronting HIV and TB. Because of West Nile, by 2001 New York State's arbovirus team had grown to twelve members.

Better education would help, too. The CDC and the United States Department of Agriculture (USDA) held a conference in November 1999 and another in January 2001 to educate public-health officials from around the country about the new West Nile threat, particularly which surveillance methods were working and how best to control mosquito populations. The CDC also set up an online system called ArboNet so that officials could easily share information over the Internet about how far the virus had spread. Congress expanded the CDC's budget so it could help states develop the infrastructure they needed—including lab facilities and personnel—to set up surveillance for West Nile in mosquitoes, birds, horses, and people. Mosquito-control efforts were also supported.

After West Nile broke out, Layton also instituted a program whereby her staff regularly mailed doctors reminders that it really was important to report suspected West Nile virus cases. The New York State health code lists seventy diseases, from *E. coli* O157 to meningitis to botulism, that physicians are required by law to report. Yet it just doesn't happen as much as it should. "I trained in upstate New York," Layton says, "and I was never told that I was legally required to report certain diseases. At hospitals, doctors tend to rely on 'other people' to do that for them—the infection control nurses, the lab directors. Many physicians are not aware of the requirements, and others just don't want to deal with it."

Layton's health department has written publications for doctors, opened a toll-free reporting phone line, and set up an advisory group to figure out how to make reporting easier. They're standardizing forms that can be used for all diseases. "But we can't just do these things once," Layton says. "Physicians come and go. The West Nile outbreak shows how important reporting is. On the date we started our investigation in 1999, we knew of four cases in the hospital. In retrospect there were fifteen other people who were already hospitalized but had not been reported. We were just lucky Dr. Asnis called when she did."

Layton also enlisted the public in the battle against West Nile, asking people to report dead birds to health authorities. So many sightings were called in that the health departments could test only a fraction of them. But as West Nile spread into each new area, the dead birds were the first indication of its arrival. As our cities and towns continue to be bombarded with infectious diseases, citizens should not wait for instructions. They should be alert to unusual die-offs. And we each can become more actively involved. We can write letters to our newspapers and call talk shows saying how important it is to be on the lookout for new diseases. We can also quiz political candidates about how much they know—or care—about public health, and let those already in office know that their constituents support spending money to protect their community from infectious threats.

In addition to expanding local and global monitoring of infectious diseases, epidemiologists and public-health officials must develop better systems for sharing their findings quickly. Outbreaks of all kinds can be minimized by better communication across the wide-ranging medical, lab, pharmaceutical, and public-health communities. That's what limited deaths from a summer 1999 outbreak that killed three women in southern California. The individuals had never crossed paths, yet they shared one fatal connection: CDC epidemiologists determined, by isolating and testing virus genes, that the women had fallen prey to a new pathogen spread in the dried feces of field mice. The scientists named it the Whitewater Arroyo virus.[16] The virus had actually been discovered, before it was known to cause disease, during a CDC field survey done by researchers looking to see if cousins of Lassa fever had established themselves in mice in this country. The epidemiologists' rapid detective work, complete with genetic sequencing and DNA fingerprinting, connected the Lassa research to the newly described Whitewater Arroyo virus, and public-health officials' fast and widespread warnings to avoid areas frequented by mice succeeded in shutting down a potential outbreak before it started.

But quick communication requires doctors, nurses, and other health-care workers who are trained to recognize potentially dangerous and suspicious occurrences, as Deborah Asnis did when Enrico Gabrielli entered her hospital's doors. Then they must actually report them to the proper authorities. The Internet has improved communication flowing from centralized experts to doctors around the country and the world, and online initiatives should be expanded. Ten years ago a doctor would have read the journals in his field and perhaps the *MMWR,* the CDC's famous *Morbidity and Mortality Weekly Report,* reprinted in general medical periodicals like the *Journal of the American Medical Association.* Beginning in the mid-1990s, the emerging-infectious-disease division at CDC began to build online networks to connect doctors, state health departments, and CDC offices. Today those networks are used to issue advisories and, to a lesser extent, to report diseases. ArboNet is one such network, now used to expedite reports when local labs find special arboviruses like West Nile; Marci Layton relies on it during each West Nile virus season. FoodNet is for food-borne outbreaks like *E. coli* O157. PulseNet allows agencies to identify multistate outbreaks of bacterial infections.

But as Michael Miller, who, as chief of CDC's epidemiology and laboratory branch, is one of the nation's leading microbiologists and epidemiologists, points out, the mere fact that resources like the Internet exist is not enough to ensure our safety. We must also utilize the resources we have in an effective manner. "Internet linking has become the cornerstone of public-health communication," Miller says. Yet, ironically, with so much information online it can be impossible to notify the entire health community of an important development. "Dermatologist networks advise dermatologists, cardiac surgeon networks advise cardiac surgeons. But when an outbreak may be under way, everybody in the health-care network needs to be informed with one voice," Miller says. "That doesn't happen."

To help, the CDC is developing a Health Alert Network that would link all public-health labs with the CDC and act as a bridge connecting doctors, hospitals, and public-health officials. Yet even if the network succeeds in linking all the specialized forums, it will then bear the burden of being the nation's single source of advice, and guidelines will always take a certain amount of time for CDC experts to issue. "CDC can be that voice," Miller says. "But that voice has to be correct, and that requires time. Once CDC makes a statement, it will affect a massive number of people who might receive treatment. For each case we have to gather data, do field research; we couldn't advise a nation on how to combat, say, a smallpox outbreak overnight." The changes being made will, however, surely lead to vast improvements.

Another vital improvement that is needed for our public disease-detection system is better testing capability. The West Nile case illustrates well that lab tests can be inconclusive or even wrong when technicians face a foe they've never seen. The state and the CDC, using antibody-based tests, first misidentified the virus as St.

Louis encephalitis. Lipkin thought it was Kunjin virus when he looked at limited stretches of the genome. Misidentification can easily happen when two organisms are related and have similar-looking surface proteins. They're like family members who resemble each other; an acquaintance approaching them on the street mistakes one brother for another, until he gets up close and recognizes each individual's distinctive eyes or nose. The misidentification problem was compounded by the fact that St. Louis encephalitis was known to be present in the United States and West Nile wasn't. The state labs didn't have any antibodies to West Nile virus, so they couldn't test for it. The CDC had the antibodies, because it had been consulted in the Romanian outbreak, but didn't test for them initially, since West Nile virus had never been seen in the United States. The CDC should have considered the possibility of something unusual, rather than sticking with the common suspects. It was also slow to connect the importance of the bird die-off, again relying on old information that these viruses infected but didn't kill birds.

Today, because of the anthrax scare and the West Nile outbreak, many states can perform tests for arboviruses as well as for unusual pathogens like anthrax and plague. "We hadn't had arboviruses in the city since yellow fever in the 1880s," Layton says, "so we could not have justified the expense of having that testing capability." Now Layton can, but her agency's abilities continue to be limited by scarce funds. Testing is a game of resources. Most local health departments don't even have their own labs. New York City does core tests only for potentially deadly pathogens known to rear their ugly heads there, such as botulism and rabies. To ease budget crunches, state and local health departments can apply to the CDC for supplemental funding for certain measures. About 90 percent of New York City's tuberculosis-lab funding comes from the CDC, and it gets assistance for programs that fight AIDS and sexually transmitted diseases.

Ironically, better surveillance capabilities put in place at the state level because of the anthrax scare were a key reason that the 2002 West Nile outbreak was confronted sooner than it would otherwise have been, especially in Louisiana, where it hit hard early in the summer. CDC director Gerberding said some of the detection problems presented by the West Nile outbreak were similar to those that health departments would face during a biological attack by an unusual organism.

Such expanded federal support, however, adds to the nation's tax burden. The alternative is privatization of labs—allowing them to be run as commercial businesses. Layton is wary of the idea, though. "Once you privatize, you're dependent on industry to maintain high levels of expertise, and to be willing to maintain ways to rapidly mobilize for emergencies. Can they afford to have such capabilities that would rarely be used? What are their financial incentives? Initially only one commercial lab could test for West Nile virus and it was charging a lot of money—$100 per test."

In addition to increased surveillance and better communication and testing systems, prevention is a key element in keeping us safe from emerging and reemerging infectious disease. Thanks to the existence of appropriate funds and the infrastructure with which to distribute them effectively, New York has been able to put in place preventive measures against West Nile virus. Workers aggressively spread larvicide to cut the mosquito population, a far safer step than spraying insecticide that can compromise people's health. Public education helps, too. In 2001 the city put posters on its fleet of garbage trucks, which stop in front of almost every home, encouraging people to eliminate any source of standing water.

The most effective prevention, of course, would be the development of more vaccines. The National Institutes of Health are supporting research to create a vaccine for West Nile as well as vaccines for other encephalitis organisms. NIH is the leading source of support for basic and applied medical research in the country, and it runs its own research organizations on its Bethesda, Maryland, campus, including the National Institute of Allergy and Infectious Diseases. In 2000, NIH issued a $3 million grant to Acambis, the same company that came through with the smallpox vaccine, to develop a West Nile vaccine. The company now has an experimental vaccine based on a live, genetically altered strain of yellow fever, called 17D, which has been used widely in yellow-fever vaccines since the 1930s. Researchers hope the new vaccine will preserve the safe and effective properties of the yellow-fever vaccine, but confer immunity to the West Nile disease. This technique was first developed by Thomas Chambers at St. Louis University, and Acambis has patented the key procedures. The same technique could create vaccines for other viruses in the same family, which cause dengue fever as well as a form of encephalitis that is endemic in Japan. The technique also has the potential to create a vaccine for the more distantly related hepatitis C virus, which, as we'll see later, has become a major health threat.

Yet despite all the potential benefits of this line of vaccines, the funds for further development are limited. NIH has said it will support the work through "phase I" clinical trials—the first of three trials, each involving a larger group of human test subjects, which the FDA requires before it will approve a drug as safe and effective. Phase I trials are run to test the safety of new products, while phase II and III trials are meant to show that the drug or vaccine works effectively. After phase I, Acambis will be on its own. The economics of vaccine creation and production are problematic, however, limiting the amount of this kind of development, and as we will explore later, a more concerted, better-funded system is one of the most important mechanisms we should put in place for the pathogen war.

Better public-health monitoring, better lab tests, and vaccines are all crucial to stopping many sorts of infectious diseases early. But they do nothing, of course, to cure the diseases that will inevitably make their way here and infect some portion of the population. Although public-health officials eventually managed to stem the 1976 Legionnaire's outbreak in Philadelphia, anywhere from 8,000 to 18,000 people continue to be infected each year, and dozens die. Microbiologists still do not know how to stop the *Legionella* bacteria. The recent outbreaks of West Nile, Lyme disease, plague, and Whitewater Arroyo virus also show that, too often, our medical and pharmaceutical communities operate from a deficit. Once a new infectious disease breaks out, they need time to identify it and study how it works before they can stop it. Many people can die in the interim. Three years after West Nile's surprise attack on New York City, there was still no antidote.

The good news is that a host of new molecular and genetic techniques are being developed that will greatly enhance our ability to create more effective drugs. As scientists sequence the full set of genes—the genomes—that make up more and more pathogens, they will gradually discover the specific genes that instruct each organism in how to infect us, reproduce, or create toxins. This information will help drug makers design medications that target those mechanisms, improving the kill rate of drugs while lessening troublesome side effects. And one genetically engineered drug could even be able to block an entire class of disease organisms.

The work is exceedingly complex, however. Finding a cure for West Nile virus, or any virus that causes the brain swelling of encephalitis, would be quite a feat. The brain is protected from viruses, bacteria, and toxins in the blood by the blood-brain barrier. Cells lining the walls of blood vessels in the brain are more tightly knit than in the rest of the circulatory system. They stand shoulder to shoulder, like the Royal Guard lined up across the gateway leading into Buckingham Palace. The guards don't budge for anyone, making it difficult for any molecule that doesn't present a proper password to cross the living wall and reach the brain cells behind it.

The blood-brain barrier saves our lives daily. Without it, all sorts of pathogens and poisons would bombard our neurons, and we would surely die. Scientists aren't certain how viruses that cause encephalitis gain entry. Whatever the mechanism, in effect, one of the guards in the shoulder-to-shoulder wall steps aside, letting the virus slip through. Unfortunately, as is so often the case in microbiology, a good defense can also be an Achilles' heel. The guards lining the blood vessel walls won't allow most drugs to enter the neuron palace, either. If a pathogen does manage to slip inside, it can wreak havoc, while the guards stop drugs from following it in and destroying it.

Once a virus gains access to the brain, it might burrow into neurons, reproduce, and destroy the cells. The brain has only one limited form of immune cells, so the virus can pretty much run free. Only as it multiplies and causes destruction do the guards outside realize hell has broken loose behind them. They then give way so that immune cells from the bloodstream can enter and fight. Evidence suggests, however, that the immune system often makes matters worse, frequently getting overexuberant in battle. Immune cells come sweeping in recklessly. The powerful enzymes the immune cells use to destroy viruses and bacteria can spill out and damage surrounding neurons, resulting in mounting inflammation of the brain as well as actual brain damage. Even if a person survives the pathogen attack, he may sustain neurological damage as a result of his own body's defense mechanisms—which may explain why West Nile victims like Enrico Gabrielli still have difficulties.

In trying to design a treatment for West Nile, researchers aren't even sure yet that the virus does indeed sneak into the brain through the blood-brain barrier. It might take some other route, by infecting nerve cells in, say, the arm after a mosquito bites. Then it could feel its way along the nerve fiber to the spinal cord and follow that up into the brain stem. The herpes viruses can do this, and herpes simplex virus is the most common cause of encephalitis in the United States.

Researchers can only experiment with drugs that they surmise might help. Ian Lipkin, the California molecular biologist who identified Marci Layton's samples, has since published data indicating that ribavirin stops the virus from replicating in specimens grown in the lab. Ribavirin has been used with mixed success to treat hepatitis C. But ribavirin doesn't cross the blood-brain barrier, either. And a test by the Rabin Medical Center in Israel of thirty-five West Nile patients there showed ribavirin had little effect. Virologist Robert Sidwell at Utah State University has since begun testing other possible drugs in lab specimens.

The bottom line for now is that there is no cure for West Nile virus, or for an increasing number of exotic infectious diseases causing outbreaks outside their places of origin. The West Nile invasion has served as a wake-up call about just how quickly an exotic disease can settle in and become indigenous. Quick and effective action in New York probably slowed the spread of the virus, but even so, a squadron of infected birds has taken it across the country and into Canada. The story emphasizes how important it will be in the future to keep alert for new diseases, which means making sure that our public-health system stays strong and up-to-date, and that it takes advantage of the latest in communication and education techniques to keep doctors and ordinary people informed. Unless we improve our ability to alert health officials and the public quickly to the emergence of a potential outbreak, we're vulnerable to any number of new outbreaks of foreign diseases.

CHAPTER FOUR

MAD COW AND CHRONIC WASTING DISEASE: *The Strange Horrors of Prions*

Not only are we under assault by exotic pathogens from afar, but we've recently discovered that we are also threatened by a strange and entirely new class of pathogen: prions.

For years, scientists didn't even believe prions existed. They dismissed the claims of a few maverick colleagues who said they had found a normal animal protein that was horribly disfigured, and was somehow causing massive brain damage. Stanley Prusiner, a professor of neurology and biochemistry at the University of California in San Francisco, led the renegades. He first identified prions in 1982, and hypothesized that they caused the sheep disease scrapie, which gradually ate holes in the animals' brains. Prusiner also proposed that prions might cause the rare, always fatal disorder known as Creutzfeldt-Jakob disease (CJD), which ruined human brain tissue, but he couldn't prove it. Skeptics said what he had really found was merely a virus that acted slowly.

Then mad cow disease hit Britain in the mid-1980s. Cattle started foaming at the mouth, acting insane, and, before long, collapsing dead in their tracks. Autopsies showed that their brains had been eaten away, as if worms had bored holes through their gray matter. Prusiner and others examined the tissue and said prions were to blame. The brain destruction in cows looked strikingly similar to that in scrapie and CJD. Researchers named the new disease bovine spongiform encephalopathy, or BSE, evoking the spongelike quality of what remained of a cow's brain after the disease ravaged it.[1] Ten years later British men and women in their twenties started dying of a strange new variant of CJD, labeled vCJD. Prusiner argued that both BSE and vCJD were caused by prions.

Yet even as Prusiner accepted the 1997 Nobel Prize for discovering prions, critics objected that malformed proteins simply could not set off a chain reaction of damage that constituted an infectious disease. Finally, in 2001, as more analysis of brains from sheep, cows, and people became available, the medical community conceded that prions were indeed causing these gruesome diseases, as well as several other rare human diseases that attacked the brain, including Gerstmann-Ströussler-Scheinker syndrome (GSS) and fatal familial insomnia (FFI). At last the medical community had to admit the awful truth: The human species now had to combat a whole new type of pathogen.

Unfortunately, scientists still have not figured out how prions actually destroy brain cells—or how to stop them. All of our classic defenses are useless. Our immune system doesn't know how to fight them. There are no vaccines and no effective drugs. Prions cannot be destroyed even with radiation. And prion diseases bring our defenses down in ways doctors have never seen and turn victims' minds into mush.

Confronting prions may demand new social thinking, too. When the British government ordered millions of cows slaughtered in the late 1980s and early 1990s, officials figured they had BSE licked. They were confident the disease would not cross the species barrier and infect humans. But it already had. By the late 1990s young Britons who had eaten contaminated beef a decade earlier began dying. Health officials in the rest of the world figured their citizens were safe; as long as people didn't eat beef from British cows, they'd avoid prions. But in 2000, human victims began appearing in other European countries.

American officials said there was no worry in the United States because the disease was confined to cattle, British imports had been banned, and the U.S. stock was clean. But in 2001, veterinarians warned that mule deer and elk in the Rocky Mountain states were dying from an uncommon disease that caused their bodies to waste away.[2] Was this just a coincidence? No. Scientists showed the game were dying from yet another prion infection, which they labeled chronic wasting disease (CWD). Health officials, flouting recent history, said they doubted humans could contract the disease, although they advised hunters to not eat the meat of animals that appeared sickly. Then, in the summer of 2002, pathologists feared that three hunters had died after doing just that.[3] Holes had been gored in the men's brains. And chronic wasting disease was spreading through deer and elk herds across the American and Canadian Great Plains.

The capacity of prions to impart disease may just be beginning. Even though the vCJD scare in Britain has been limited to fewer than 200 human victims so far, we may just be witnessing the early stages. And chronic wasting disease may have already set in motion a latent human outbreak in the United States, the effects of which we might not see for years. Researchers have no idea where prions will turn up next.

The most frightening, and confounding, aspect of prions is that they are so stealthy. Victims have no idea they are sick until they start behaving oddly. By then, holes are already forming in their brains. And even when the disease is full-blown, doctors find it hard to diagnose. The one reliable test is a brain biopsy, which can only be done after a patient dies. Unless individuals and our doctors are alert to the hidden clues of the disease's early stages, an outbreak could spread widely before it is even recognized. Detecting those clues is much easier said than done, as the story of Pamela Beyless shows.

Pam grew up in the bucolic English countryside, a few miles from the tiny old village of Queniborough, 100 miles north of London.[4] She delighted in the small white Tudor homes trimmed in black, some with centuries-old thatched roofs, lined neatly in a row on the lone main street that gave way to open pastureland. Despite their modest income, Pam's parents, Arthur and June, made sure to put good meals on the table for Pam and her two brothers. The family's doctor, Philip Monk, always declared the kids healthy, and Arthur, a stocky man with a broad smile, intended them to stay that way. June, a cheery talker, loved the small-town life. She purchased fresh bread and vegetables from the local baker and grocer whose shops were on the main street, and bought her meat in bulk from the town butcher, Ian Bramley. In the traditional way, Bramley killed and dressed cows, sheep, and lambs he acquired from the neighborhood farmers. June had to pay a bit more for his choice cuts than she would have at the general market, but she thought the quality was worth the price. She bought her processed meat, like meat pies, at the market to save money.

Pam didn't necessarily love schoolwork, but by fifteen she did love the boys at Soar Valley Comprehensive School. She'd gossip with her friend Cindy about what would become of them if they never got married. Cindy would say that Pam had no worries: she was tall, so pretty—especially her brown hair—and so outgoing. Sure enough, boys called her for dates. Arthur, always the steady provider, did the taxiing so he could keep an eye on them. He trusted Pam, but not the rest of the world.

As soon as Pam graduated, she told her parents she and her boyfriend, Andy, intended to move into their own place in Southampton, four hours away on England's southern coast. Pam was just seventeen, and her parents were anxious about the idea. But Pam made friends there quickly with two girls, Jo and Trudy. They'd go out clubbing, drinking and dancing till dawn at the nightclubs, and eat late-night kebabs from the street vendors—ground lamb on pita bread—or get burgers at McDonald's. Before long, Pam and Andy bought a house, and were soon engaged. Pam was content with her job as a clerk at a real-estate agency. "All I really want," she'd tell Trudy, "is to get married and have babies."

Arthur and June moved a few miles from Queniborough to the village of Glenfield, closer to the big town of Leicester, where Arthur started a new job delivering milk to people's homes. Pam and Andy went to visit them for Easter 1994. On their second night there, Arthur was up in his bedroom when he heard Pam shouting. She was downstairs crying when Arthur got there. She had found a note from Andy on the settee. All it said was "Pam, we're finished." She hadn't seen it coming.

Though Pam grieved for months, at least she still had good

friends. Trudy and her boyfriend moved into the Southampton house and helped pay part of the household bills. Trudy also persuaded Pam to go out more again, and soon the fun-loving Pam was back. On her twenty-first birthday, in October 1995, her new boyfriend, Dave, and Trudy, Jo, and other friends took Pam out for a big pub dinner. Pam was seated in the middle, grinning from ear to ear.

The respite was short-lived, however. Trudy and her boyfriend soon moved out to buy a house of their own, and Pam could no longer afford her home. She sold it and ended up renting a bedroom at Trudy's place. She felt disappointed that she was so far from having a family of her own. She also started worrying about her job, because she'd found herself making silly mistakes lately.

All the upheaval in her life made Pam feel more alone. Realizing that, in June 1996 she sent Arthur a glowing Father's Day card. He was delighted that she had thought of him, but when he called to catch up shortly thereafter, he felt that Pam sounded surprisingly withdrawn. He guessed she was still depressed over losing her engagement and her house, and he thought a visit home might help. When he suggested to Pam that she drive up to see them in Leicester, though, she said the trip was too long. He thought perhaps she couldn't afford the gasoline, which was peaking at seventy-one pence a liter, the equivalent of $3.50 a gallon in the United States. "We'll pay your petrol," he told her. But still she declined.

Pam didn't reveal to her father the real reason she wasn't keen on driving that far. Recently her friend Sarah had asked Pam to teach her to drive. Pam would drive them in her black Renault stick-shift out of town to an open area with straight roads, and hand the wheel over to Sarah. On their return one Saturday, though, Pam kept stalling the car. For some strange reason she kept forgetting to put her foot on the clutch when she came to a stop. She laughed it off, but stalled again and again. The truth was she didn't trust herself to drive to Leicester safely.

Arthur and June were troubled by Pam's resistance to visiting, so the next Tuesday they drove down to Southampton and brought Pam back to Leicester for a break. When Pam's boyfriend, Dave, arrived the following Sunday to pick her up, her parents thought Pam seemed strangely distant toward him. Arthur figured she was just down about having to go back to Southampton and face her frustrations again.

Pam's relationship with Dave was still casual, and she didn't see him much in the coming weeks. She began keeping to herself, holing up in her room most nights and on the weekends. She wouldn't go clubbing with Trudy. One warm Thursday night that August, as Trudy headed upstairs, she peeked through Pam's door at the top of the stairway and saw Pam sitting in her robe on her bed, bent forward. She had taken a bath and was somberly combing her hair, which was coming out in small bunches. "Why is your hair falling out?" Trudy asked, surprised at the sight. "Are you taking something?" she questioned, worried that her friend might be using drugs of some sort.

"I'm not taking anything," Pam protested. She slumped in despair and blurted out, "I can't concentrate anymore. At work, they try to teach me something, but I can't understand it." She was deeply troubled.

"Aw, don't worry," Trudy said, trying to snap Pam out of her mood. "If you lose your job, we'll look after you."

Pam just smirked in reply, and then suddenly exclaimed, with a seriousness that startled Trudy, "Oh, I *am* going to lose my job. I don't understand it anymore."

The next morning Pam came down for breakfast and chatted to Trudy as if their awkward conversation from the night before had never happened.

A week later, when Trudy was mopping the kitchen floor, Pam entered the room in a crying fit. "I can't understand why I can't think straight," she sobbed. As days passed and Pam continued to have such mood swings, Trudy's suspicions grew. *You hear it so much, about drugs,* Trudy thought. *And you never know what's really going on with somebody.* Trudy determined to keep a closer eye on Pam.

Several days later she found Pam in her bedroom, which was strewn with clutter. "Are you going to clean up your room?" Trudy asked, half joking.

"What?" Pam said, looking confused.

"I said, are you going to clean your room up?"

"What do you mean?" Pam answered. Trudy walked away, shaking her head. What was happening to her friend?

The next morning Pam happened upon Trudy in their small backyard. "My God, my room is a mess," she announced. "I've really got to clean it up." Trudy was stunned. It seemed that at times Pam was perfectly normal, but at other times she was in a strange, detached state. Her behavior was like nothing Trudy had ever encountered.

At the end of that month, in late August, Pam lost her coordination, just like that. She was at work and began staggering down a hallway, as if she were drunk. Her manager saw her and asked suspiciously, "What's the matter?" Pam said she didn't know. The manager sent Pam to the company doctor, who suggested she visit her own doctor, which she did later that day. He found nothing physically wrong, and told Pam she was depressed and under a lot of stress. He suggested she take a little time off.

"No way," Trudy responded to the diagnosis that evening. "You're not depressed. Something else is wrong." She was trying to stir some fight in her friend. In addition to Pam's odd moods and now this coordination problem, Pam hadn't been eating well, either. The mother of her friend Jo noticed that Pam looked as if she was losing weight, and was so concerned she rang up Arthur and June. "Something isn't right with Pam," Jo's mom said. She suggested it might help if Pam went home to Leicester for a while.

Arthur and June did come to get Pam, who agreed to spend some time with them. Once they got her home, they quickly realized their daughter had a serious problem. "What's this business with the walking?" Arthur asked when he saw Pam swerve out the front door one morning. Pam just shrugged. He wondered if so much stress had built up in Pam that she couldn't control herself. *Stress can do awful things to you,* he thought. *It can tighten your muscles; you can get infection after infection because it weakens you.* He had had a bit of trouble walking himself, and had terrific pains down the fronts of his legs, when he had been locked in a long, tense standoff with his boss in his previous job.

Arthur and June figured time would heal Pam's distress. They fixed a simple bedroom for her upstairs. Pam would descend to the living room in the morning. She'd sit on the green floral settee, her back to the front window, and watch the console TV across the room. On the damp September mornings, Arthur would light the gas-fired charcoal heater inserted into the old fireplace beside the television. Pam would stare blankly at the small flames.

Each day Pam seemed more sullen. She'd want the living-room curtains closed to shut out the light, and more and more she wanted Arthur or June in the room with her. They tried to persuade her to go outside, but the idea seemed to frighten her. One day, when her old school friend Cindy came for a visit, Pam just sat quietly on the settee, looking a bit nervous. Cindy would start a conversation, but Pam could not stick with any line of thought. She'd go quiet in midsentence, look around the living room, look back at Cindy, and then her eyes would just wander off.

Within a month of coming home, Pam could not get up off the settee without Arthur's or June's help. Arthur decided they had to do something. He brought Pam to their new family doctor. The physician examined Pam and ran a few blood tests, but could not find anything wrong. One thing was clear to Arthur, though—he wasn't letting Pam go back to Southampton anytime soon. Some unusual illness must be affecting her, and he was now going to push the doctor to find out what it was.

Though Pam was often distant and confused in those days, she did have moments of clarity. When her brothers drove her back to Trudy's house to pick up her belongings, she saw her friend Jo and confessed: "I know something's going wrong, Jo, because I have to think about walking." She said she didn't have any pain, but with each step she would have to visualize how to lift her leg and where to place her foot. Every movement of her legs, arms, and neck required that kind of concentration, every minute of the day.

Arthur kept pushing the family doctor for answers. The doctor told Arthur he should admit Pam to the psychiatric ward at Glenfield Hospital, just up the road. He reasoned that because he couldn't find any complications in her physical exam or blood tests, perhaps she had some kind of psychological condition. For a week the Glenfield doctors asked her numerous questions, gave her standard psychiatric tests, then declared that her problem was not psychiatric; it had to be physical.

Stymied, the family doctor figured there was only one other possibility—some sort of neurological disorder. He referred Pam to a Dr. Abbott, a neurologist at the Leicester Royal Infirmary. After reviewing her case, Abbott admitted Pam for observation.

When the doctors examined Pam, she kept claiming that she didn't really feel sick. They performed a number of neurological tests, but could never diagnose anything definitively. Even after weeks, they had little idea what was going on. Each week Arthur would ask Abbott, "Is it MS, doctor? Is it stroke?" And he'd reply, "No." He just kept saying they'd continue looking for an answer.

As is true for so many people who suddenly find themselves thrust into the arcane world of medicine, Arthur and June put their faith in the specialists. They were in shock over Pam's strange condition, and they held fast that the doctors would solve the mystery.

One morning, after Pam had been in the infirmary for several weeks, Arthur was dropping off milk at a customer's front door when he happened to look down at the newspaper lying on the stoop. The headline facing him read: "Young girl, twenty-four, dies of CJD." He picked up the paper and read it right there. It described an intractable affliction called Creutzfeldt-Jakob disease. The story of the young girl's deterioration described all the symptoms Pam had: lethargy, loss of coordination, periods of incoherence.

The global medical community had become familiar with CJD in the 1950s. It was a rare form of encephalitis that developed very slowly, but would eventually ruin the brain. Autopsies revealed holes in the brain tissue. In a few instances the cases seemed to be hereditary. But most cases popped up at random; they didn't cluster, and they didn't pass from person to person. So researchers termed the illness sporadic CJD. It killed several dozen people a year in Britain and several hundred annually in the United States.

Doctors had been perplexed about what caused CJD until Prusiner advanced his controversial theories in the 1980s.[5] He had become interested in brain diseases when he was a neurology resident at the University of California, San Francisco, hospital. There he had seen a patient die of CJD. He was impressed that a disease could destroy a person's brain while leaving the body intact. After ten years of research, Prusiner became convinced the agent was a revolutionary protein he called a prion.

Prions were like no other pathogen ever observed. All other pathogens known up to that time, and all life-forms in general, use fundamental nucleic acid molecules called DNA or RNA as the blueprint for making new copies of themselves. Prions are different. They don't have DNA or RNA. They are in no way living creatures, and they don't reproduce. They are truly bizarre.

As Prusiner and others learned more, they discovered that the prion was a variant of a normal protein produced in our bodies. That protein is called PrP. A prion was made of the same protein material as PrP, but was shaped differently. Normally proteins, which start out as a string of amino acids, fold into certain established geometric forms, each with its own function. But the prion was a mess, misfolded in a dysfunctional tangle. For reasons scientists didn't understand, a prion could somehow cause neighboring PrP proteins to disfigure themselves, too. Those deviants in turn corrupted others, creating a very slow but gradual chain reaction of deformity. The malformed PrP proteins would clump together and start to kill off neurons in the brain, creating holes in the tissue.

Because prions were composed of normal protein material, standard blood tests would not find any indicator that something was wrong. That was why doctors had no definitive evidence of CJD until they sliced open a dead victim's brain and found the telltale holes. It was impossible to tell if someone was infected by prions for sure until that person was dead. And looking back at neurological drugs that doctors had given victims, none seemed to have had any affect.

Terrified that his daughter might have this awful, incurable disease, Arthur rushed to Pam's room at the infirmary. When Abbott came in to examine her, Arthur immediately asked, "Does Pam's problem have to do with CJD?"

Abbott said he was looking at it, but he told Arthur that Pam didn't fit the pattern. CJD victims were almost always in their fifties or older. Arthur was relieved, but remained worried. The doctors still had no other idea of what was wrong. Pam would spend six weeks under the probing eyes of the Leicester Royal Infirmary's team of specialists, longer than Arthur and June would have ever imagined. But even after all that time, the doctors were unable to come up with a diagnosis. Although Pam's physical condition was continuing to deteriorate, they discharged her in late November 1996 without being able to tell Arthur and June what to do next. Arthur found that incredible, but sometimes doctors simply don't know what else to try. They had found no definable illness, and could not justify keeping Pam in the hospital any longer.

The British medical community was following a process similar to that pursued in the United States and most Western countries: examine the patient, run tests, compare the results to those for known diseases, and zero in on the likely culprit by process of elimination. When all the possible culprits are eliminated, however, doctors must simply release patients. In Pam's case, at that time, few people outside of the National CJD Surveillance Center set up in Edinburgh had an inkling about this new prion disease that was emerging.[6] Pam's case didn't fit the profile of CJD itself, and vCJD was still below the medical community's radar.

By the end of 1996 Britain's national health authorities were only beginning to put some pieces together. They had ten suspicious deaths, where autopsies indicated that there might be some "new variant CJD," abbreviated as vCJD. The early symptoms of the ten alleged victims tended to be psychiatric, including depression, anxiety, and withdrawal. But these were so common to many other illnesses that they could not indicate vCJD reliably at all. Physical symptoms like difficulty in walking seemed to follow a few months later, but these were also symptoms of more common neurological diseases, which doctors would want to rule out first.

The truth is that diagnosing new infectious diseases is fraught with difficulties. Misdiagnoses are often made, and some doctors would rather be noncommittal than wrong. Other doctors simply don't know a new disease is coming along, so many cases that should be caught go undiagnosed. If the patients are lucky, their immune systems find a way to beat the mysterious invader. If not, they can end up injured or even dead.

After Pam was released, her condition continued to worsen steadily. By early December she could no longer control a fork or spoon, even if she focused intently on how to move her wrist and fingers. Trudy called one evening, and Arthur had to pick up the phone and hand it to Pam because she couldn't do it herself. Pam struggled mightily to get words out as Arthur stood beside her. "When . . . you . . . coming . . . for . . . a . . . visit?" Pam asked. "I . . . want . . . to . . . see . . . you."

Not long after, Trudy arrived, and she was alarmed by her friend's condition. It was midday, but Pam was still upstairs in her bedroom. Her boyfriend, Dave, who was keeping in touch with her, was also visiting. He was in her room with her when Trudy walked in. Pam was lying on top of her bed quilt in rumpled clothes. Dave was sitting beside her. When Pam saw Trudy, she clutched at Dave in fear, scowling at Trudy. Horrified, Trudy watched as Pam writhed around on the bed, obviously scared and unable to talk. Trudy tried to speak to her as if she were the same old Pam, not this strange, gaunt creature with sunken eyes, sallow skin, and thinning hair, behaving like someone who was mentally ill. "Hello, how ya doing?" she said in her usual singsong fashion. But Pam simply couldn't communicate. Trudy stayed at the house two hours, but Pam never moved from her bed or spoke.

Several days later, Pam's old school friend Cindy stopped by for a visit. June was manually exercising Pam's legs, as the doctors had suggested, to improve her circulation, since she rarely walked now. Cindy and June reminisced out loud, in front of Pam on the settee, about how Cindy and Pam would play pranks when they were kids. Pam stared at Cindy for a long time, as if she knew who Cindy was and what Cindy and her mother were talking about, but still she could not speak. It seemed to Cindy that Pam was trapped inside herself. Cindy kept an upbeat attitude while she was with Pam, but when she got into her car to leave, she broke down in tears. It was obvious that Pam was seriously ill, and it was maddening that nobody could figure out what was wrong.

One of the strangest things about Pam's condition was that she could still focus at odd times. In past years she had loved visiting her brother and sister-in-law, Angela. Through all the haze of the last months, she seemed to have picked up on the fact that Angela was now expecting a baby. One afternoon on the settee, the otherwise silent Pam suddenly said, "I'm going to be an auntie." A week later Angela came to visit, and sat across the living room from Pam. After several minutes Pam slid herself off the chair and crawled across the floor to Angela. She laid her head across Angela's pregnant belly, clearly listening for the baby inside.

Because Dr. Abbott and others at the Leicester Royal Infirmary had not been able to diagnose Pam, Abbott told Arthur that he wanted to send Pam to some colleagues at the National Hospital for Neurological Diseases and Neurosurgery in London. Perhaps neuroscientists there, with more funds and expertise, could crack Pam's case. In February 1997, now nine months after Pam's troubles had begun at Trudy's house, Arthur and June took their 22-year-old daughter on the three-hour drive to the city. Arthur parked the car. He got out onto the noisy, bustling street corner and looked up at the imposing redbrick hospital standing squarely against the gray sky.

Pam was admitted under several doctors and a Professor Macdonald. Arthur wondered if these doctors would bring up CJD, the only thing he had heard of that sounded at all like what Pam might have. But they never used the term with Arthur. They did seem particularly worried about Pam's memory loss, which had become profound somewhat quickly. Arthur and June had a hard time coping with the thought that Pam might have no memory of her childhood or her identity. So, when the doctors weren't around, Arthur would test her. One morning, after Pam had been lying still and silent in the hospital bed, Arthur pulled up a chair and said, "Pam, remember when we went on holiday with your friend Zoe? You put her up on her horse, then you got on your horse and took off like a shot?"

Pam suddenly started to laugh, though she continued to look off into space. Arthur was thrilled that Pam still had some semblance of herself somewhere inside.

The doctors were familiar with CJD, and with the preliminary findings the national health authorities had released on the ten vCJD cases. The new variant CJD differed from sporadic CJD in that it struck younger people, and seemed to cause uncommon brain-wave patterns on EEG tests, but these were not conclusive. The best the medical team could do was perform various tests to rule out other rare neurological conditions as they gradually moved toward a conclusion.

After about two weeks, Arthur and June finally got the word from Professor Macdonald that Pam indeed had vCJD. They understood the gravity of the condition. When Arthur asked how long Pam might live, Macdonald said she would likely die by July.

Arthur didn't know what to do. Britain's medical laws would allow Pam to stay in the hospital for care, even though there was nothing doctors could do for her. He had done his own research on CJD since he had seen that newspaper article, and he had never found any mention of a cure. Trying to see past their own sense of doom, Arthur and June decided it would be best to bring Pam home, care for her there, and hope the doctors could come up with some kind of treatment. He was ready to wheel Pam out when the medical team said they wanted to move Pam to St. Mary's Hospital, across town, for a tonsil biopsy. Perhaps that would show something. Arthur was ready to try anything, so he agreed.

What Arthur didn't know was that Britain's leading prion researcher, John Collinge, was affiliated with St. Mary's Hospital. Collinge had been intensely researching sporadic and variant CJD, and thought he had evidence that prions might collect not only in the brain but in the lymph nodes. If Collinge could harvest some lymphocyte cells from a patient's tonsils and examine them under a microscope, he thought he might be able to find signs of prions. And perhaps analyzing the prions might lead to some clue about how to stop them. It was a long shot, but at this point it was the only shot.

St. Mary's was legendary, one of the oldest hospitals in London, and it looked it—four floors of crumbling brown brick arranged in a closed circle, like a war-torn fortress. It had housed legions of Britons during scourges of a century ago: polio, smallpox, tuberculosis. It also was slowly being decommissioned as London built more modern medical facilities nearby. The hospital staff laid Pam on a gurney and wheeled her down a hallway with Arthur and June in tow. They pushed the gurney through a set of big, old swinging wooden doors into a hectic, wide-open ward. The space was enormous and long, with dark, creaky wood floorboards. Two rows of beds stretched endlessly. Down each row, scores of men and women lay interspersed nearly shoulder to shoulder, coughing, hacking, sweating, bleeding. Arthur couldn't imagine what diseases these people had. The heavy double doors banged loudly, echoing across the ancient hall, each time a person entered or left.

Several times, Pam's tonsil biopsy was scheduled, then postponed. On the fifth day of waiting, June said to Arthur, "The only way to get help, I've learned by being here, is to shout your head off." So Arthur did. The next morning Pam had the procedure, and the following day Arthur and June brought her home. They had had enough of hospitals, and St. Mary's had been one of the worst experiences of their lives. The only reason Arthur had put Pam through it was that during the first day, one of the nurses had said the test could perhaps show a sign of some other disease that might be treatable.

Now, like many other cursed patients with terminal afflictions around the world, Pam was in medical limbo. The doctors had no next step. If she had grown up in the United States, she probably would not even have been allowed to stay at a hospital, because she had no treatable condition and therefore wasn't eligible for any kind of long-term care. All her parents could do was try to keep her comfortable and wait for a breakthrough.

Pam was deteriorating more rapidly now. She needed help going to the toilet. She couldn't bathe. Suddenly one morning she lost the ability to swallow. Arthur and June began pureeing Pam's food and spoon-feeding her as if she were a baby. All Pam could do was gulp it. She choked often.

The Beylesses didn't hear any results from St. Mary's for weeks, despite Arthur's constant phone calls begging for information. Finally, in April, Macdonald told the couple that the test confirmed that Pam had new variant CJD and that it was indeed fatal. He asked the couple if they had any questions. Crushed, Arthur and June were speechless.

It seemed to Arthur that to the doctors, researchers, and simply everyone else in the world, Pam was now a statistic. She had a terminal illness, like someone with inoperable cancer or advanced Alzheimer's. It was time for her to die. They offered no more information, and no hope. Arthur and June were determined to give their daughter the very best care they could. They had lost faith that the doctors would continue to attend to her case, much less tell them how she could have contracted this gruesome disease or how to help her.

Pam rarely talked anymore. Occasionally she'd mouth a silent word. But late in April, on a Saturday, while Pam was propped up on the settee, she suddenly said to June, "Well, I'm sorry, Mum." June was shocked. She quickly responded in a tempered tone: "What have you got to be sorry for, Pam?" But Pam said no more. The next evening, while Pam was sitting up in bed, June fed her dinner—spaghetti, potatoes, and tea—all of which Pam could swallow without choking, though it took her a long time. The food eventually went cold, so June took it downstairs to warm it again. When June returned, she saw that Pam had fallen backward onto the bed. Pam, staring at the ceiling, said, "Mum, will I be saved?" June thought Pam must have had an overwhelming sense that she was going to die.

Pam never spoke again. She never really moved again, either. Yet she hung on. The summer of 1997 came and went. Pam could no longer control her bowels. She did seem to have an awareness of her surroundings; she seemed to like the aromatherapist—Cindy's aunt—who came out of sympathy to massage her skin with oils to improve her circulation. Arthur kept pushing the hospital to prescribe some sort of stimulation. One doctor set up three visits at an experimental room run by a woman from Scandinavia. It was bathed in mood lighting and music. During one visit the woman picked up a cobweblike strand of lace and shook it across Pam's face. Pam jerked her head to the side—she hadn't moved for months! Arthur tried it again at home and Pam jerked her head again; the thing aggravated her. Arthur was thrilled.

It took fifteen months after Pam had left the hellish St. Mary's ward before Arthur could obtain a special wheelchair. Nobody would take responsibility for anything having to do with Pam. Here it was, May 1998, and finally he could take Pam outside. He'd wheel her up and down the narrow street, stopping momentarily at the neighbors'. He thought, *If Pam can't walk, can't talk, and she's stuck in one room day after day, her mind will surely deteriorate. And if I keep Pam busy, it might encourage her to think positively.* Though Pam's head didn't move, when she was outside her eyes roamed everywhere. One afternoon, when Arthur pushed Pam to the top of Chestnut Road, he noticed a big, lovely cat to his far left. Suddenly Pam turned her whole head and torso and looked right at it.

Such rare signs gave the struggling couple the strength to battle the drudgery, day after day. For all Arthur knew, Pam might have been fully aware of what was happening around her. Arthur and June found other threads of hope, sometimes in awful places. At one point in the spring, Pam was having breathing problems. Her doctors admitted her into the Leicester infirmary. The nurses put her in a ward next to another young patient, Stacy Robinson, who also had vCJD. Some weeks later, back home, Arthur heard from Stacy's parents that Stacy had died. Stacy's symptoms had begun later than Pam's, yet here was Pam, still hanging on. Maybe she was different from the other victims.

As summer turned to fall, however, their last hopes faded. One warm afternoon in early October, Arthur got a call on his cell phone while he was still at work. It was June. "Come home quick," she said. Arthur drove directly there. Pam was in bed, having real trouble breathing. They called for an ambulance, and Arthur carried her downstairs. When he put her in her chair, her eyes suddenly flew wide open. Then she closed them. The freakish gesture scared Arthur. The ambulance arrived and brought Pam to Glenfield Hospital.

Given her grave situation, Pam had to remain in the hospital.

Arthur would stay with her twenty-four hours, then June would take a turn. On Sunday, October 11, 1998, the staff gave Pam an oxygen mask so moisture could be introduced into her hardening lungs. That evening, Arthur started to move Pam in her bed to make her more comfortable. Just then, a bedside machine with a little wire that led to a clip on Pam's pinky started to bleep. It was the machine that monitored the oxygen absorption in Pam's body. Arthur thought he had jostled the mask on Pam's face. But the machine kept getting louder, which meant Pam's oxygen was decreasing fast.

Arthur hurried out to find a nurse in the hall. "Come look at Pam quick," he said. "Oh, the machine has to settle down," she answered. "No, it's had time," insisted Arthur, who was accustomed to the machine's habits. Rather than argue, he rushed down the hall and grabbed a doctor. The physician came running while nurses followed. It was now about 7:30 P.M. The beeping was getting louder and louder, the oxygen value lower and lower. Arthur, on one side of the bed, and June, on the other, could see that Pam wasn't breathing. "Breathe, Pam, breathe!" Arthur shouted. June saw Pam's eyes fly wide open, just as they had when Arthur put Pam in her chair. Then Pam died.

Nobody could help her. The doctor, the roomful of nurses—they all just stood there.

Pamela Beyless was the twenty-seventh person to die of vCJD as of October 1998. She was twenty-four years old. Although the medical community was beginning to accept that the disease was caused by prions, it still had no idea where those prions were coming from. Were the misshapen proteins manufactured by the victims' own bodies? Were people picking up the prions in some way? There were perhaps twenty hypotheses made by various researchers about what might cause the strange diseases: ferret bites, blood transfusions, immunizations, even cornea transplants.

The mystery of how people were contracting CJD and vCJD was finally solved in part due to the ongoing persistence of none other than Arthur Beyless, and to the determined sleuthing of a pair of local public-health examiners.

Pam's funeral was a small, quiet affair. Cindy came. Trudy could not bring herself to attend. Arthur and June mourned for several weeks. But as the days wore on, Arthur began to reflect a bit more critically. During Pam's various hospital stays, he and June had gotten to know Stacy Robinson's parents, as well as the parents of another vCJD patient. Both families also lived near little Queniborough. He was hearing more on the news about other young victims of this "new variant" of CJD in England. Something must be happening to cause the disease, yet no one in the health community or the government was owning up to it. Surely, he thought, the medical community must know more than it was saying.

If some kind of cover-up, or just plain incompetence, had contributed to Pam's death or suffering, Arthur could not stand the injustice. Furthermore, he thought, more parents could lose their daughters and sons if the word about vCJD did not get out widely. But what could he do? He had a clever idea. He figured the newspapers would know how to investigate the situation. And a news story would provoke other people to pressure the health system and the government for answers. He phoned the *Leicester Mercury* and told a reporter there that three young people in the immediate area had all died of new variant CJD, yet no doctor or government health official had acknowledged there might be some local cause.

The *Mercury* ran a story in December 1998. It landed on the desk of Philip Monk, who had been the Beylesses' family doctor in Queniborough when Pam was a little girl.[7] Monk was now the chief public-health examiner at the countywide Leicestershire Health Authority. The article quoted Arthur Beyless, his former patient, and said it was odd that with only twenty-seven vCJD deaths nationwide, three were connected to the same little village of only 3,000 people. Could there be a local health problem?

Monk knew he'd soon get calls from politicians, physicians, and citizens wanting answers. So he contacted Bob Will, a neurologist at Edinburgh University who ran the fledgling National CJD Surveillance Center, which tracked cases. Will met Monk in Leicester and the two concluded that the three instances did not necessarily constitute a "cluster." Although three victims living that close together might sound unusual, statistically it was within reason. Thus Monk, who had plenty of the usual health concerns to pursue, did not begin an investigation. But in May 2000 a nearby doctor phoned Monk and asked for advice on diagnosing a patient he thought might have vCJD. Then, a month later, Will called Monk and said he had received word of yet another human death in Leicestershire from vCJD. There were now five men and women between the ages of nineteen and thirty-five, out of only seventy-four known cases nationwide, who had died or were dying of vCJD, all on Monk's turf.

The next morning, Monk was just going to start reviewing the cases when he was called away to a nearby school to deal with head lice. His assistant examiner, Gerry Bryant, did some calculating on the back of an envelope while he was gone. The surprising result alarmed her; statistically, five people out of 3,000 with CJD was altogether different from three people. She called Monk on his cell phone to tell him that the risk of getting vCJD in Leicestershire was about fifty times greater than the national average. They had a real problem.

Monk and Bryant gave themselves a crash course in prion diseases. They plowed through past investigations of people who had died from sporadic CJD and vCJD. They also burrowed into the scientific literature about BSE, hoping to find clues as to how that infectious prion agent was transmitted among cattle and what its incubation period might be—information they might be able to apply to the human CJD cases.

In August 2000 the examiners began to interview relatives of the victims. Step by step, Monk and Bryant asked about a host of standard activities that were known to cause disease, as well as the novel proposals from researchers. One by one they ruled out possible links among the victims. They didn't all have ferret bites. As kids, they had not all eaten commercial baby food. They had different dentists. They weren't all agricultural workers. They received vaccines from different manufacturers. They didn't all have body piercings, or drink from the same water supply. The interviews did provide a reliable time frame: 1980 to 1991 was the crucial era during which the victims would have lived close to one another.

More and more factors fell by the wayside. Finally, a single trait seemed to link all five victims: they ate beef, and their main source of beef in the 1980s had been local butchers, not supermarkets.

Monk and Bryant now had to learn all they could about livestock raising, slaughtering, and butchering practices. What could it be about beef that might actually cause a prion disease in people? By September they were talking to people involved in all the steps in the meat supply chain. It seemed odd—in fact, suspicious—that the symptoms of cattle with BSE resembled the symptoms of people with vCJD: the loss of muscle control, the loss of mind. Was it possible for humans to succumb to a cow disease? And if so, why didn't they die from it way back then?

By this time the national media had gotten wind that an investigation of vCJD was under way in Leicester. And a few scientists like Prusiner and Collinge were saying that it might be possible for people to get vCJD from ingesting infected animal tissue. They cited notes made in the 1960s by British researchers who were investigating jungle diseases. Natives in certain tribes in Papua New Guinea, like the Fore, were cannibals, and ate the flesh and brains of their deceased tribesmen as part of the funeral ritual. A common cause of death among the Fore was a disease they called *kuru,* a word that meant "to tremble uncontrollably." Descriptions from the researchers' autopsies of victims' brains seemed to match the description of brains ravaged by BSE and vCJD. As missionaries progressively convinced natives to stop cannibalism, kuru stopped, too. Monk's phone began ringing. National newspaper reporters were calling. Was there a vCJD outbreak? Was it linked to BSE? TV reporters streamed into the little town. The story had all the right elements for high drama: a rare killer disease, fear that any individual could be the next victim, and all of it breaking loose in this otherwise idyllic village.

Monk and Bryant continued to interview more people. After four months they felt they had found the smoking gun and could prove it. Monk first reported the findings, in private, to the victims' relatives. Then he announced that the Leicestershire Public Health Authority would hold an open meeting on March 21, 2001, at the Queniborough Rugby Club, which had the biggest auditorium in the area.

It was a freezing cold morning when Monk drove up the bumpy drive to the club. Falling snow was covering the dark grass. Monk walked to the podium at the edge of the stage. He faced a sea of people, and was surrounded by cameras and journalists from Britain and around the world. Monk explained calmly that five people in Leicestershire had died from variant CJD, which the media was now calling the human form of mad cow disease. A nine-month investigation of the area's food supply had concluded that the victims had died from a common cause. During their childhood and adolescence they had eaten meat purchased at one of the two butcher shops on the main street in Queniborough. The meat had routinely been contaminated with bits of cow brain that had unknowingly harbored BSE.

Gruesome as it would be, Monk had to go into detail. At large slaughterhouses, he explained, cattle were stunned, then their heads were lopped off. Their bodies were put on conveyor machines, then butchered. But at small abattoirs like the local butchers, only one cow would be prepared at any given time. The butcher killed the animal with a gunlike device that drove a steel rod, called a pithing rod, into the cow's brain. The butcher would retract the pithing rod and carve up the body.

Although the pithing rod killed the cow mercifully fast, it sometimes caused some brain matter to splatter from the head. Bits could fall onto the animal's tissue as the butcher carved out portions of meat. Knives used to split the heads for disposal might later be used to cut the meat, and the splitting was sometimes done on the same big tabletops where the meat was trimmed. And the pithing rod itself was sometimes used from one animal to the next without cleaning. Each of these activities could spread tiny bits of diseased brain matter onto portions of retail meat.

Scientists, Monk continued, had also shown that BSE and vCJD were both caused by an unusual protein called a prion. They were now theorizing that if a person ate these proteins in beef, the pathogen could possibly survive the person's digestive system, be absorbed into the blood, and slowly eat away the brain, ruining the victim's coordination, speech, and mind.

Monk stressed that the butchers had not done anything illegal, or even wrong. The butchers' actions were common practice back then. Indeed, it was the best of butchers who were still processing their own cows. The government hadn't labeled cow brain as risky material until 1990, owing to the rise of BSE. That's when butchering practices had changed, and both local and national meat producers became much more careful to keep brain material from contacting anything else. He also said that scientists did not know how long the disease could lie dormant in a person's brain before beginning its evil attack. Because four of the five victims had left Queniborough by 1991, it seemed the disease could smolder for years. All Monk could say with certainty was that it was highly unlikely for meat to be contaminated now.

The next morning the world's newspapers and TV news shows were filled with stories about mad cow disease killing people. Monk quickly lost track of the coverage, however. One week after the big meeting, a huge outbreak of tuberculosis began at a school only a stone's throw from his Health Authority office. He had to turn his full attention to it—yet another killer infectious disease.

The wider medical community, however, took up the cause of determining how many people might have been infected with vCJD, but to this day it remains a murky exercise. No one knows how many butchers across Britain would have engaged in practices similar to those used by Queniborough's butchers. Nor do they know whether brains from infected cattle that were slaughtered at abattoirs big and small were safely disposed of. Were they instead mixed with other beef by-products to make the filling for meat pies and sausages, sold nationwide? Were they ground up into ruminant and added as a protein supplement to feed other cattle—a common slaughterhouse practice? If cows were eating ground brains, they subsequently could have developed the disease and passed it on to people when they were killed. This might mean that multitudes of people have vCJD right now, but have not yet developed symptoms.

People carrying the deadly disease may have spread it, too. If they donated blood, they could have passed on prions to others who received blood transfusions. In November 2001, the British news show *Panorama* aired an investigative report that disclosed that eight vCJD victims had given blood during their adult lives. The show also claimed that prions could adhere to surgical utensils and survive the standard hospital cleaning regimen known as autoclaving, in which utensils are blasted with steam to kill germs. If true, prions could perhaps be transmitted on surgical forceps and clamps from patient to patient, day after day. The *Panorama* show even featured a microbiologist who said the National Health Department knew about this possibility, but was keeping it quiet so as to not cause panic. The United States soon outlawed blood donations from people who had spent more than six months in Britain at any time between 1980 and 1996. In 2002 the American Red Cross, working with the Canadian-British firm ProMetic, began trying to develop a test for prions in the blood supply.[8]

New-variant CJD prions might well have been transmitted in other ways. Gelatin made from cow hides and bones is used worldwide as a base for women's cosmetics, for drug capsules and pill coatings, and, in a cruel irony, as a serum for vaccines against other infectious diseases. It is highly unlikely that prions can survive the physical and chemical processes involved in creating such products, yet the possibility cannot be ruled out. In 2000 the U.S. Food and Drug Administration discovered that five vaccine makers were still using cow blood, meat broth, and calf serum from countries where BSE was known to exist. The products were part of nine vaccines, including those for children's polio, diphtheria, and tetanus, as well as the military's anthrax vaccine.

Experts still argue about when BSE began in cows, and that data is necessary for estimating the incubation period of vCJD. The British government officially recognized BSE as a disease in 1988. Because the designation had been in the works for about two years, and because the first human vCJD cases were identified in 1996, some forecasters say the incubation period for vCJD is about ten years. But Monk and others had uncovered stories about cows dying after having "gone mad" as far back as 1980. That would mean the incubation period could be as long as sixteen years.

All the camps agree that exposure stopped by 1996; although the British government banned the use of bovine brain and spinal cord from all human and bovine food products late in 1990, it did not ban the use of ground ruminant in cattle feed until 1996.

Whether vCJD's incubation is ten or sixteen years leads to widely diverging estimates of the number of possible deaths still to come.[9] Simon Cousens, a statistical epidemiologist at the London School of Hygiene and Tropical Medicine who had conferred with Monk early on about the Leicester cluster, and Peter Smith, the school's head of infectious and tropical disease, assume a ten-year incubation period. Their model indicates that the number of cases should have peaked by 2002, and will now trail off. "It's hard to say there would be more than a few thousand cases altogether," says Smith, who is also the British government's lead adviser on spongiform encephalopathy. As a young researcher, Smith had trekked through the Ugandan jungle in the 1960s in search of viruses that might cause cancer, and his own research back then showed that kuru's incubation was about ten years, too. Yet if many Britons did indeed carry the disease without getting sick but passed it on through blood donations or through surgical instruments, vCJD could still spread widely. "That is a worry," Smith notes.

Other experts say the London School's numbers are too conservative. Neil Ferguson and Roy Anderson, leading epidemiologists at Imperial College in London, say known cases of BSE go back to 1982, when the disease was dismissed as scrapie in cattle. Ferguson also says that farmers widely underreported mad cow deaths, so their herds were not seen as suspect by meat buyers. Since BSE's incubation period in cattle is only five years, BSE would have been widespread and entering the human food chain in quantity before the 1988 government classification. Ferguson estimates as many as 50,000 to 100,000 people could die from vCJD.

The only accurate indicator of how many people may die comes from the annual death toll: three in 1995, ten in 1996, ten in 1997, eighteen in 1998, fifteen in 1999, twenty-eight in 2000, twenty in

2001, seventeen in 2002. So far the numbers do not seem to be increasing significantly, and thus we can hope that lower estimates will prevail.

Who is most susceptible seems a bit clearer: children. Most of the victims die in their twenties. If the incubation period is at least ten years, they must have been children when they were exposed. If adults were as susceptible, numerous older people would have died by now. For the victims' parents, this demographic conclusion has been especially hard to swallow. "If you go by the incubation period," Arthur says, "then Pam must have gotten it from the food we fed her as a child."

But why just certain children? So far, scientists have discovered that the victims shared a common genetic trait: their genes that encode their PrP proteins carry two copies of a mutation at a certain location. This trait is shared by 40 percent of the British population. Thus not everyone who eats infected meat would become sick, another hopeful sign that deaths from vCJD will not escalate out of control.

As more of the story of vCJD's rise becomes filled in, Arthur Beyless becomes troubled that doctors might have known earlier that Pam had vCJD and could have done something to forestall her death, in case treatments were found. He feels the doctors had decided that since vCJD was incurable and Pam was declining, she'd be better off dead. Months after Pam died, Arthur asked county administrators to look into her case, and he notes that they later found the Leicestershire County Council Social Services Department guilty of "maladministration." The administrators also found that Pam had never been properly discharged from the first hospital, the Leicester Royal Infirmary, and because of that, subsequent hospitals and social services were reluctant to further her treatment. A later report by the Leicestershire Area Health Independent Review indicated that the other hospitals had improperly discharged Pam, too. Through an attorney, Arthur confronted the Leicester Royal Infirmary's executives, who ended up admitting their wrongdoing and offering him £3,000 ($4,800) in compensation. In March 2001, Arthur initiated an investigation of the

National Hospital's administration, but more than a year later it was still bogged down in report writing.

The case of mad cow disease, including the transgressions Arthur uncovered, speaks volumes about the difficulties of quickly identifying the onset of a new infectious disease, and about the limitations of the public-health system. The British government slaughtered millions of cattle in the early 1990s to stop BSE, and by 1996 it declared vCJD a new disease, but the connection remained obscure. By March 1997, British labs were using a technique called "strain typing" to show that the makeup of vCJD was very similar to that of BSE. Tests in 2000 showed that lab mice infected with BSE and vCJD exhibited the same prion misfolding in their brains. Independent critics who reviewed the timing of the findings said that government health officials and food inspectors were inept in not having suspected a link much sooner. Others claimed the beef industry had persuaded the government to wait out the vCJD cases; there were still fewer than 100 in the year 2000, so why put people in a panic about beef? All the uncertainty raised the stakes of answering the most controversial question of the day: How many people might have been infected with the mad-cow prions? By the end of 2001, several vCJD victims had appeared in Ireland and France.

The slowness in defining or announcing vCJD in the United Kingdom also may have slowed international realization that prion diseases in herd animals might cause human disease. In 1967, American scientists identified a BSE-like illness, chronic wasting disease (CWD), in deer and elk. They had no reason to believe the disease could threaten people, so they never sounded a loud public-health alarm. If in the 1990s U.S. officials suspected a link between eating diseased animals and human illness, they might have ordered deer and elk in areas of infection to be slaughtered, to stop the disease's spread. As recently as the late 1990s, the disease seemed to be confined to animals in the area where Colorado, Nebraska, and Wyoming meet. By 2002, however, CWD had spread dramatically, to fifteen states and two Canadian provinces.

If hunters have been eating contaminated deer and elk, and if that can initiate a form of CJD, and if the incubation period is ten to sixteen years, as it seems to be from BSE, the United States could be on the verge of its own human outbreak of "mad game" disease. Indeed, late in 2002, evidence was mounting that there seemed to be an unusual number of vCJD–like deaths among outdoorsmen or people who eat venison. Three men who regularly shared game feasts at the home of Gary Waterhouse, an avid hunter in Chetek, Wisconsin, all died in the past decade of the disease.[10] Waterhouse died in 1993 at age sixty-six, and James Botts died in 1999 at age fifty-five. The third partner died of a similar brain disease that doctors now think could have been CJD, although it was not diagnosed as such at the time.

CJD is divided into subgroups depending on how it is acquired. When a potential new version of the disease is discovered, it is referred to as CJD until it can be better defined. "Sporadic CJD" is the most common type, affecting one in a million people worldwide each year—200 to 300 in the United States. Inherited or "familial" CJD results from specific mutations in the PrP gene and is very rare. "Variant CJD" is the human form of BSE. All these afflictions cause spongy-looking brains, but the number and patterns of the holes can distinguish them. Since 1996 there have been four American cases of CJD in people younger than thirty. Three of the four were venison-eaters, and sporadic or familial CJD in people so young is extremely rare.[11] Statistics indicate a minuscule likelihood that Waterhouse and his friends, or the four young adults, contracted CJD by chance. Still, these cases remain officially classified as sporadic CJD, although the CDC is now investigating them. CWD had been initially dismissed as the cause of death because details in the pattern of damage in the victims' brains looked like sporadic CJD and not so much like vCJD. But since a transfer of CWD to humans would be a new disease, no one can say what the brain should look like.

The August 2002 death of Gary Padgham, an elk hunter in Bozeman, Montana, could shed more light.[12] At age fifty, he had had the stumbling gait, impaired vision, and memory loss typical of CJD for the prior ten months. He was diagnosed with CJD in

Seattle, and his body was sent to Prusiner's lab in San Francisco for autopsy. The results are not publicly available because of confidentiality issues, but they could provide conclusive evidence of the link.

There is growing concern among medical researchers that the inability or unwillingness of public officials to draw the connection between CWD and these deaths is similar to the initial assurances by the British government that people needn't be concerned about mad cow disease spreading to people. The CDC's expert, Ermias Belay, says there is no conclusive evidence that CWD has spread to people. But he also says it can't be ruled out. Meanwhile, many locales in Midwestern and Western states that derive great revenues from hunting are downplaying the threat. After CWD was detected in animals in Wisconsin in 2001, state officials wanted to kill off all the deer in a 374-square-mile region where the disease was found,[13] but they met objections from landowners and hunters. Health officials did at least begin advising hunters to avoid cutting into eyes, brain, spinal cord, spleen, tonsils, and lymph nodes of any deer or elk, and to soak dressing knives for an hour in a 50/50 solution of bleach and water.[14] Realization of the potential danger does seem to be making its way into the public consciousness. A poll taken by the *Milwaukee Journal Sentinel* in September 2002 showed that more than 60 percent of people who had eaten Wisconsin venison in the last five years said they were unlikely to eat venison now. Unfortunately, with rare and unusual diseases like CJD it will be difficult to establish CWD as the source of disease until more victims are identified, and that could take a few more years. In the meantime it may be prudent to avoid eating venison from states in which the infection has been detected unless the deer has been checked for CWD. Hunters and suppliers should freeze the meat until test results show that it is safe.

Variant CJD could be a one-of-a-kind aberration. Or it could be a scary sign of more prion diseases to come. More research is needed, and fast, given the imposing potential of related threats like CWD. In 2002 there still was no commercial test for vCJD, much less a vaccine or cure. We know much less about prions than we do about bacteria and viruses. By the time a person has been diagnosed based on neurological malfunctions like those that Pam Beyless displayed, her brain is already badly damaged, and that damage is virtually impossible to reverse. The tonsil biopsy that Pam endured is the closest thing we have to a test, but it requires a surgical procedure—and that only if the person still has her tonsils. It is hardly a test that can be administered routinely. And it's still unclear whether prions always collect in the tonsils anyway.

Scientists have also developed an experimental test for CWD that can be performed on thin slices of brain and lymph node tissue taken from dead, infected animals. In the summer of 2002, Wisconsin health officials, who are taking CWD seriously, outfitted a lab in Madison and expected to test 50,000 specimens taken from game killed during the upcoming hunting season.

Much better would be a blood or urine test that would detect prions in the very early stage. Since millions of people would want to be tested, and farmers would want to test millions of cows, a drug company could make billions of dollars manufacturing such a test. Sure enough, by 2001 more than twenty American and European companies were racing to find that test. Yet the first advance came from an Israeli scientist who had studied scrapie for years. In July 2001, Ruth Gabizon, a neurologist at Hadassah University Hospital in Jerusalem, announced she had discovered a unique substance, a tiny, never-before-seen protein, that was present in minute amounts in the urine of people with a rare, hereditary form of CJD.[15] It also appeared in the urine of cows with BSE. Gabizon figured that prions could accumulate in victims because the body's enzymes could not break them down. Chances were, then, that some would occasionally be filtered out by the kidneys and be expelled in urine.

Several prion labs around the world worked in 2002 to try to validate Gabizon's results. If they hold up, after trials and regulatory approvals they might lead to the first test that can be administered to live people and cattle. Gabizon and a few investors started a spinoff company named PrioSense to develop such tests.

Designing a cure will be more difficult. First we must understand much more about how prions cajole normal proteins to misfold. It is unlikely this work will happen at pharmaceutical companies. The harsh reality, says the London School's Peter Smith, is that "neither the pharmaceutical industry nor the government will invest millions of pounds [sterling] to develop a cure when there have been only a hundred cases" of vCJD.

So finding a cure may fall to academics. New evidence from several labs on both sides of the Atlantic suggests that a misguided immune system might be an unwitting accomplice in prion diseases. Research in mice suggests that once prions are ingested, they move to lymphoid organs like the spleen. While there, they begin to cause PrP proteins on lymphocytes and macrophages to refold badly. From there the expanding clan of prions probably moves along peripheral nerves to the spinal cord, and migrates up to the brain, where they find a treasure trove of PrP they can continue to convert. The studies further propose that immune lymphocytes do not identify prions as intruders, so they don't attack. Perhaps because the amino acid sequence of prions is the same as that for healthy PrP protein, the immune system does not detect the misfolding.

Researchers hypothesize, however, that as misfolding prions collect into aggregates in the brain, they do attract the attention of microglial cells, the brain's simple immune cells.[16] Pathologists find high concentrations of these cells when autopsying vCJD victims. The microglial cells may mistake the PrP clumps for some sort of bacteria. Unfortunately, the microglial cells, in a desperate effort to eradicate the prions, may accidentally kill neurons, creating the spongy holes. The immune system may unintentionally destroy the brain while trying to protect it. The same microglial cells apparently also react rashly to the presence of certain proteins in the brains of Alzheimer's patients, though to a much lesser extent.

The immune system could redeem itself someday, if it is really to blame. Perhaps some as yet undiscovered immune trick actually does get rid of prions before they reach the brain. That would explain why so many people who ate tainted beef did not develop the disease. The trick might be lacking in people who succumb.

One promising man-made treatment against prion diseases may come from an experimental strategy being tested on Alzheimer's patients. Very preliminary data reported by Frank Heppner and colleagues at University Hospital in Zurich suggests that in mice, man-made antibodies can break up the Alzheimer's plaques and prevent disease.[17] This experiment is one step toward better understanding how to eliminate protein aggregates that could underlie more common degenerative brain diseases such as Alzheimer's, Parkinson's, and amyotrophic lateral sclerosis (ALS).

Collinge, Britain's leading prion researcher, is looking to genetics for clues. In 2001 he became director of the St. Mary's Prion Clinic. He got the green light from St. Mary's and £12.4 million ($19.8 million) to build a state-of-the-art research lab on two floors of the tall, modern Institute of Neurology building that looms over the old, squat, red-brick National Hospital for Neurological Diseases, where Pam stayed before she was moved across town to the St. Mary's ward. The lab includes high-tech DNA sequencers, ultralow-temperature freezers, and special microscopes. Half of one floor is built to biosafety level 3 (BSL-3) containment standards. Collinge also wooed several of Europe's leading prion experts to join him. In 2002 he began what would be a long search to see if vCJD victims shared certain genes that were not present in the general population, which might help explain why they had developed prion disease.

The other route to a cure, of course, is to experiment with drugs on people, and a contest is under way between Prusiner and Collinge to do just that. Prusiner's lab took the early lead in the summer of 2001 when it began treating a patient with the most unlikely of compounds—an obsolete drug used in World War II for malaria.

Any drug that would be able to combat prions would have to cross the brain-blood barrier that keeps most substances out of the brain. Only thirteen classes of drugs can get past the guard cells there. Early in 2001, Carsten Korth in Prusiner's lab began testing each class on prion-infected cells harvested from lab mice. Only one class seemed to affect the prions. So Korth tested dozens of drugs in that class. By spring 2002 he had found two that inhibited the formation of prions. The more effective compound was quinacrine, little used today but administered widely to U.S. soldiers in the 1940s to treat malaria.[18]

Soon after Korth had his mice results, twenty-year-old Rachel Forber showed up on the lab's doorstep. She arrived from Merseyside, England. According to British news reports, she had been diagnosed with vCJD in June and was given a year to live. Her father, Stephen, had scoured the Internet and determined that Prusiner was the world's leading expert on prion diseases. He called Prusiner cold to ask if the expert could help his daughter. Prusiner's office seemed sympathetic, but said it had no kind of treatment. Stephen called and called, until one of Prusiner's colleagues disclosed that they were trying experimental drugs on mice. Desperate, Stephen wanted Prusiner to try them on Rachel, but the colleague said the lab did not have federal approval to begin a drug trial, and that quinacrine was not cleared for use in CJD cases. Finally, the colleague agreed that if the FDA would allow it, the lab would test the unlicensed drug on Rachel. The FDA said okay under a controversial doctrine called "compassionate use," whereby doctors can use an unapproved drug or one licensed for another use to treat a dying patient in the hope that it will have some effect. Stephen and Rachel flew to San Francisco immediately.[19]

Stephen and Rachel arrived at Prusiner's lab on July 4. Rachel didn't know where she was. Her speech was slurred. Like Pam, she could no longer feed herself. She began taking pills three times a day. After six days on quinacrine, her skin turned yellow—an expected side effect. But after only three weeks, Rachel's coordination and state of mind seemed to improve.

Stephen leaked snippets about Rachel's surprising progress to the British tabloids. With a large supply of pills, father and daughter returned home to England. Although Rachel's improvement in neurological symptoms continued, within a few weeks it became clear that Rachel's liver could not withstand the toxicity of the heavy drug doses. Rachel would have to go off quinacrine or she would die of organ failure. Rachel relapsed quickly. Four weeks later she died.

Rachel's story provided hope that effective drug treatments

might be developed. The media coverage prompted American and British regulators late in 2001 to find a lab in each country to perform a clinical trial of quinacrine, as well as chlorpromazine, the less effective drug that had been tried in mice. Stanley Prusiner's lab got the nod in the United States. John Collinge's lab would do the UK trial at St. Mary's. Each team designed protocols independently, but they turned out similar: Test patients would be admitted to the hospital for two weeks, then take the drugs as outpatients for up to six months. The news in 2003 wasn't encouraging, though. Further experience with quinacrine demonstrated only temporary improvement even in those who could tolerate it.[20]

Prusiner now plans to screen thousands of quinacrine analogs for a more potent drug and is developing an artificial antibody that recognizes the odd form of PrP.[21] He also is investigating why the normal and aberrant PrP proteins fold differently, and thinks the answer may relate to how sugars attach to PrP. Prusiner is quite interested in CWD, too, and is researching whether prions from infected deer or elk can cause human PrP to change into the prion shape. All the work is preliminary. If it leads anywhere in animal tests, it will then likely take five years or longer to move into human trials and gain regulatory approval for public use.

The sudden emergence of prions offers important lessons about fighting infectious diseases. First, the medical community must be on higher alert when patients present strange symptoms. It might have been excusable for Pam's friend Trudy to think Pam's odd behavior might be due to drug abuse, but the first doctors Pam saw dismissed her lethargy and loss of mental acuity as the results of stress. If she had been that way for months, perhaps their notion would be acceptable. But Pam's symptoms came on quickly and were a marked change from how the formerly fun-loving and energetic young woman had always behaved and felt. Doctors are taught to suspect the obvious first, but when that's as far as they go, they've lost their edge. How can we have any confidence that physicians will nip infectious outbreaks in the bud, if they simply stick with the usual suspects?

Pam's story also emphasizes how important it is for patients to push back when they do not believe they are being treated well enough. Pam's condition made that impossible for her, so Arthur took charge. He pushed the hospitals to investigate CJD after reading the newspaper story about another victim. And yet when doctors seemed reticent, he probably could have asked even more questions, though he was perhaps so overcome with grief about Pam's plight that he didn't have the strength to fight. It was an outrage that he later had to ask and ask for help with Pam's ongoing home care. But if Arthur hadn't continued to confront the medical community, they would have written off Pam entirely. If you are a patient, or the loved one of a patient, and you are not getting enough information or help, demand attention. Do your own research. Then ask for the seemingly outrageous. That's what Stephen Forber did, and it gave his daughter Rachel a shot at surviving.

Arthur's actions after Pam's death pose larger lessons for all of us who are concerned about the rise of infectious diseases. Arthur simply could have turned his back on vCJD. But in Pam's death he found a mission: to uncover why the British medical community and health authorities were so slow to react to a new infectious threat. He started networking with other families of victims. As a group they began to press the authorities for answers. Having learned a painful lesson about how passive the establishment could be, Arthur turned to the media to ratchet up the pressure to find the cause of this new disease, and it worked. Anger over Pam's death prompted him to demand answers. And while it wouldn't have helped in Pam's case, patients and their loved ones should demand answers long before their diseases progress too far. You could be saving yourself, and scores of others, from new diseases.

Whether or not the government covered up the possible link between bad beef and human disease is still being debated. But the lesson here is clear, too: the government has got to come clean with the public, as soon as it suspects possible problems. Infectious diseases are too aggressive to be swept under the rug. Health officials and food industry leaders must stop treating the public like children; the only excuse they ever give about keeping a lid on a possible new disease is that they don't want to create panic. That's nonsense. Britain's leaders were obviously too slow to confront BSE and vCJD. The question for American leaders now is, will they react just as slowly to the spread of chronic wasting disease?

Government must also fund more research into emerging infectious diseases, and scientists should yell loudly if they have trouble getting funds. Prusiner defined prions in 1982 and soon proposed them as the cause of several fatal animal and human illnesses, but even his own peers doubted him. Prusiner won the Nobel Prize in 1997, yet the FDA, and similar agencies in the United Kingdom, didn't fund a single human drug test until 2001, after a hundred Britons had already died. It has taken twenty years merely to start to attack this sinister pathogen.

The prion saga also calls attention to a lesson that will loom large in the rest of this book: We can never underestimate how devious infectious diseases can be. We cannot, for one moment, think we have the upper hand. We clearly are not taking prions seriously enough. Variant CJD, and CWD, could be just the tip of an iceberg. Prions' capacity to impose disease may just be beginning. What happened to Pam is unlike what happens in any other illness. Even though the vCJD scare has been limited to fewer than 150 victims, we may just be witnessing the very early stages of the prions' transformation into a raging threat.

We have no idea where it may turn up next.

CHAPTER FIVE

E. COLI GONE BAD: *Runaway Bacterial Evolution*

We should be concerned not only about threats from exotic foreign diseases and newly emerging ones. The threat from everyday pathogens, all around us right here at home, is at least as great. They are evolving rapidly, and they can strike at any time.

It was a warm Friday evening in South Milwaukee. Three-year-old Brianna Kriefall loved the summer weather. She had been outside all day, playing with her Barbie doll, reading Dr. Seuss books, watering the flowers in her family's yard with her toy watering can. By dinnertime, all her play had made her hungry. Her mother, Connie, and father, Doug, decided to take Brianna and her two-year-old brother, Chad, to the local Sizzler steakhouse for dinner. It was a buffet-style restaurant like thousands of others across the country, where scores of people parade past salad and dessert bars and scoop mounds of food from metal and plastic bins onto their plates. The Kriefalls thought the kids would have a great time.

Brianna's blue eyes opened wide as she walked alongside the children's buffet. She loved the cheese, cantaloupe, and watermelon, and chatted happily as she sat down to gobble up her treats. Connie and Doug were proud of her because she ate so well. Chad didn't eat much. They told him he'd have to eat more when he got home.

The next morning Brianna complained of a stomachache, and then she came down with diarrhea.[1] Before long she began to vomit, and her diarrhea became bloody. Her condition quickly became so severe that Connie and Doug drove her to Children's

Hospital and hurried her into the emergency room. The doctors asked Connie and Doug all sorts of questions; stomachache, vomiting, even diarrhea, could signal any number of illnesses. But the bloody discharge was troubling. It could mean that toxins were in her intestines, indicating that she might have an infectious disease. A nurse drew some of Brianna's blood for testing. Connie and Doug stood by, worried but trying not to show it as they comforted their exhausted and sobbing daughter.

Brianna's bloody diarrhea was getting worse, and the doctors concluded that an infection must already be raging. They told Connie and Doug they should admit Brianna so the hospital staff could watch her closely. It was now Saturday afternoon.

Brianna worsened overnight; her abdomen cramped and contracted wildly. Lab tests showed Brianna was likely being attacked by the strain of bacteria known as *E. coli* O157. It multiplies fast and churns out toxins at a staggering rate, especially dangerous in very young children. As the toxins invaded Brianna's intestines, they triggered waves of painful contractions—her body's desperate attempts to expel the poisons.

As the toxins also rose in her bloodstream, her kidneys processed her blood faster and faster, trying in vain to siphon off poison. But by Sunday morning, less than two days after she fell ill, Brianna's kidneys, attacked by the toxin, were so taxed that they stopped functioning. She was now in a highly critical condition. Nurses hurried Brianna to the intensive-care unit, where a doctor immediately inserted a catheter into a vein in her arm, which would send her blood through a tube into a dialysis machine. It would attempt to filter the toxins mechanically, sending her cleansed blood back into her system through another catheter in her other arm.

For the next four days Connie and Doug stood next to the throbbing machine in disbelief, trying their best to cheer up Brianna when she was awake. But she slept more and more. The toxins began invading her liver, lungs, and heart, and her body started shutting down. Doctors informed the petrified parents that Brianna was battling a dangerous condition known as hemolytic uremic syndrome—toxin-induced blood clotting that leads to kidney failure. Connie and Doug were stunned to learn that there was no medication that could help her. No antibiotics work against the variant *E. coli* O157. The best doctors could do was try to counteract the awful symptoms while her immune system tried to kill the bacteria. But soon Brianna's lungs failed, and the doctors put her on a respirator. Her condition was now dire. She fought for several more days, but on Friday, six days after she had arrived at the emergency room, Brianna died.

Brianna Kriefall's death is a tragedy that could be replayed on any given day in any city or town across the country. *E. coli* O157 is lurking all around us, and the chances are considerable that at some point this genetically altered pathogen will make it onto your dinner plate or those of your children. Health officials never know how bad a given outbreak will be. They have to move fast if they hope to find the source and stop the spread.

While Connie and Doug watched helplessly as their daughter's life slipped away, health investigators in Milwaukee had been scrambling. The day before Brianna arrived at the emergency room, Mary Rotar, the hospital's infection control nurse, was already suspicious. Several other children had been admitted with symptoms similar to Brianna's. It was not unusual to see one or two children with bloody diarrhea at any given time, but when Brianna was admitted, Rotar realized she now had six. Lab reports had not yet confirmed that any of them had *E. coli* O157, but she wasn't going to wait. She asked each child's parents, as nonchalantly as she could, what their children had eaten lately. Then she called the Milwaukee Health Department. "I have a cluster of children," she told a nurse there. "The parents all mentioned dining out recently."

Health department nurses provided Rotar with questionnaires that helped her identify a possible common site of exposure: a local Sizzler restaurant. Health Commissioner Seth Foldy immediately tracked down Paul Biedrzycki, the department's director of disease control, and called him back from his vacation. Foldy then sent an inspection team to the restaurant, where they noted food-handling errors and found that employees were suffering from diarrhea. Over the weekend the lab reports from the children had confirmed that they were infected with O157, and the health department staff began interviewing parents. Sure enough, they corroborated that they had all recently dined at the same Sizzler.

On Monday, Biedrzycki held a press conference to inform the local media that the health department was looking into a cluster of food poisoning cases that appeared to be linked within the past ten days to the Sizzler restaurant on West Layton Avenue in Milwaukee. He asked them to publicize a hotline phone number that could be called by people who thought they might have been sickened. By Tuesday, Biedrzycki had received 1,800 phone calls. "I got every kind of gastrointestinal illness imaginable," he recalls. "Many had nothing to do with this case. But sorting through them, it appeared that a number did have possible ties."

On Wednesday, Foldy told Sizzler's owners that his department needed to investigate the restaurant. The owners volunteered to shut down, and inspectors swarmed the place. They put bits of food from the food bars into little vials and bags. They swabbed utensils, food-processing equipment, and countertops, and rushed all the samples to public-health labs. They also interviewed the restaurant's cooks, servers, and cashiers about how food was stored, prepared, and served. Biedrzycki even rented a hotel room, where a constant parade of employees came to give stool samples, to see if they were carrying the infection, too. The investigators soon discovered that some employees moved from jobs as cook to server to cashier on a given day; that employees worked even if they were sick; and that some Spanish-speaking employees may not have understood their training, provided in English, about proper food-handling procedures. To Biedrzycki, everything added up to one hypothesis: "The opportunity for cross-contamination was great." Somehow, *E. coli* O157 was being spread around.

By Friday, the day Brianna died, Biedrzycki had determined that twenty-one people with lab-confirmed cases of *E. coli* O157 had eaten at the Layton Avenue Sizzler from July 14 to July 21. Fifteen were children, seven of whom were being treated at Children's

Hospital, and three of whom, including Brianna, had developed hemolytic uremic syndrome.

On July 31, 2000, the Monday following Brianna's death, Biedrzycki informed both the victims and the media that the department had a prime suspect: a much higher proportion of ill people had eaten watermelon from the salad bar. Inspectors had found that the fruit bowl had been replenished each day without discarding older fruit. But it was not clear how the watermelon might have become contaminated with *E. coli* O157.

There was one strong clue, yet it was also a point of confusion. The state's lab personnel had extracted *E. coli* O157 bacteria from victims' stool samples, regrown it in a lab dish, and tested it genetically. Its DNA matched the DNA from an O157 strain found in an unopened package of raw sirloin tips stored in the restaurant.

But Brianna had not eaten meat. Neither had a few other victims. Furthermore, if the restaurant had cooked its meat thoroughly, that would have likely killed the bacteria. Biedrzycki's staff reviewed what employees had told them about how food was prepared. Then it struck them: The employees might have cut the raw, infected meat in the kitchen, then used the same knives and cutting boards to cut watermelon for the food bars. The health department staff called up several people who worked in the kitchen, who verified that this was sometimes done. That small slipup would be enough to start an outbreak; it takes only ten microscopic *E. coli* O157 bacteria to start a toxin factory in a person's gut. Once the infected watermelon was set out on the food bars, it got mixed into other foods, spreading the contamination.

In the weeks that followed, Biedrzycki concluded that more than sixty people had a verifiable infection from eating at Sizzler, and many others had also likely been made ill. Brianna was the only victim to die. The two other children who contracted hemolytic uremic syndrome managed to survive, but they could end up with long-term kidney problems. Biedrzycki told the media the outbreak was one of the state's biggest. The key to stopping it "was that first phone call from Mary Rotar," Biedrzycki says. "She felt comfortable giving us her opinion, without lab results, because there is a friendly relationship among hospitals and our health department. Given all the infectious diseases around, it's critical to have a cooperative network of astute people."

The bizarre thing about *E. coli* O157 is that it's a variant strain of another *E. coli* bacteria that is perfectly harmless. Billions of bacteria make our gut their home. We know little about them, and most seem benign, but even the harmless ones can be reservoirs of nasty genes. Friendly *E. coli* lives in the guts of various mammals, notably people and cows. It has evolved to take advantage of the warmth and nutrients the gut provides. It multiplies in such numbers that it crowds out many pathogenic bacteria that might try to establish a foothold in this cozy space. *E. coli* also produces various compounds that kill off other bacteria.

Bacteria are among the most ancient life-forms, and they are everywhere, even in your mouth and on your fingers right now. They perform all kinds of helpful tasks: They break down dead plants, animals, and sewage, converting waste into new soil to sustain the chain of life. But bacteria also cause many diseases, including strep throat, pneumonia, staph infection, food poisoning, tetanus, whooping cough, plague, syphilis, and leprosy. They thrive in our bodies, circulating in our bloodstream, collecting in our intestines, our lungs, or any soft tissue they manage to find, and may produce toxins as the by-product of their aggressive lives. Bacteria reproduce simply by dividing in two, and some do so at an astonishing rate, spawning more than 16 million copies in a single day. Certain pathogenic *E. coli* strains mutate 1,000 times faster than the harmless strains. Our immune systems can fight back against many of the pathogenic strains, and antibiotics can destroy most others. But not all.

In recent years certain bacteria have been turning more virulent, and some are evolving at terrific speed. *E. coli* has also performed the disturbing trick of picking up sinister genes from other types of deadly bacteria, creating the O157 strain. Once O157 infects your intestines, if your immune system doesn't quickly defeat it, you will soon be running to the bathroom with diarrhea. By the second day the diarrhea will be constant and severe—your body's desperate attempt to expel the rising tide of poison—and it will continue for an entire week. The whole while, your intestines will contract irregularly in painful cramps. No antibiotic will reliably help you.

If you are a healthy adult, your immune system still has a fighting chance, despite the agony. But for an adult whose immune system has been weakened by illness or age, or for a young child, the *E. coli* O157 toxins can overwhelm the body's natural defenses, poisoning organs and the blood supply, plunging the person down a slope to potential death. Most hospitalizations from O157 are for severe bloody diarrhea, but about 5 percent progress to the deadly hemolytic uremic syndrome, which kills 5 to 10 percent of its victims.

Antibiotics not only do not help, but may actually make the pathogen more aggressive. Investigators made this troubling discovery in the late 1990s, when they found that children who were fighting hemolytic uremic syndrome and were given antibiotics were much more likely to die than children who didn't receive medication.[2] They theorized that in some people, when *E. coli* O157 cells see antibiotics coming, they respond by flooding the victim's body with toxin. The massive outpouring of poison shuts down the victim's kidneys and destroys small blood vessels. But then, in August 2002, scientists at the University of Wisconsin determined that only certain antibiotics, given at certain stages of an infection, might have this effect.[3] They think other antibiotics administered at select times might help. The bottom line is that doctors still don't know exactly how to treat victims who are under attack.

Each year in the United States, 73,000 people fall ill to *E. coli* O157; 2,000 of them are hospitalized with kidney trouble; and sixty die—most of them children. In 2000, the year Brianna died, more than sixty other outbreaks occurred nationwide.[4]

The truly scary aspect of *E. coli* O157 is that colonies of the bacteria may be growing in any given package of meat you buy for a weekend barbecue, a meal you order at your favorite restaurant, or a beverage you buy at your local coffee shop. The first loud *E. coli* alarm sounded in 1992, when more than 600 people were struck with severe diarrhea and intestinal cramps after eating undercooked hamburgers at Jack-in-the-Box fast-food joints in Washington State. Forty-five individuals had to fend off hemolytic uremic syndrome, and three toddlers died. At first the medical community was unclear that an outbreak was under way. Doctors in a variety of towns across the region were reporting that patients, especially children, had bright red stools. The number of reports mounted quickly, and so widely that state and CDC health officials began to suspect tainted food was being sold by a large food chain with numerous outlets in many locations. Tests on stool samples revealed *E. coli* O157.

First identified by scientists in Oregon in 1982, the deadly strain has since spread far afield, and it can pop up at any time. Stories can be found widely in the daily newspapers. In October 1996, Randy Shaffer's experience was detailed in the *Seattle Times*. Randy had wandered innocently into his favorite coffeehouse in Seattle. He bought a latte for himself and a special apple juice for his two-year-old daughter, Taylor, who was holding his hand. Several days later, Taylor doubled over with cramps and diarrhea. Randy brought her to her pediatrician, who had blood tests done, and when the results indicated *E. coli* O157, the doctor sent her to the local hospital, where Taylor's kidneys began to fail. Randy watched helplessly as doctors connected her to a dialysis machine. In her case the dialysis saved her life. But as Taylor matures, the damage done to her kidneys could prevent them from effectively clearing even normal toxins that are produced during routine digestion and exercise, putting her at risk to become a regular dialysis patient or even require a kidney transplant later in life. She could also end up with liver disease or heart damage. Investigators ultimately showed that the unpasteurized apple juice, produced by a natural foods supplier, was to blame. More than seventy other people suffered severe symptoms from the juice at the time, and one Colorado girl died.

E. coli O157 can spread in many ways. In June 1998, scores of citizens and visitors in the tiny resort town of Alpine, Wyoming, started having diarrhea. Nineteen were soon hospitalized. CDC investigators, who asked victims what they had eaten and drunk in prior days, determined that the town's tap water had spread the organism. Strong winter storms had toppled fences erected to keep animals out of the natural surface spring that feeds the town's water supply. If even one animal had waded into the spring and left contaminated feces there, the pathogen could have reproduced and eventually trickled out of every kitchen faucet in town.

In February 2000, eleven Japanese people became ill after eating charcoal-broiled burgers at three Hungry Tiger fast-food joints in Tokyo. The culprit turned out to be *E. coli* O157 in frozen hamburger patties imported from the United States. The source was a cruel irony because Hungry Tiger had begun importing patties in 1996 after more than 10,000 Japanese were sickened by a nationwide outbreak of *E. coli* O157 tied to domestic beef.

In June 2000, *E. coli* O157 poisoned eighteen Cub Scouts and Boy Scouts at a jamboree held at an agricultural fairground in Turiff, Scotland. Investigators determined that the organism had been multiplying in sheep and cattle there, whose feces lay on the dusty ground. The boys kicked up the contaminated dust, sending *E. coli* O157 organisms into the air and into their own food and water buckets.

After ten children became ill at a water park in Atlanta, CDC investigators determined that *E. coli* O157 microbes had spread from bits of feces from a sick kid who swam in one of the pools. Children at day-care centers can burp up bacteria onto toys that other kids then put in their mouths. *E. coli* O157 also crops up in fruits and vegetables, such as lettuce, that have been fertilized with contaminated manure or washed with contaminated water. Of particular concern is imported produce, because many countries have lax food-safety standards. Nonetheless, U.S. imports are rising; 38 percent of fruits and 12 percent of vegetables eaten by Americans are now imported, according to the U.S. Food and Drug Administration.

But out of the alarming variety of sources, food that is contaminated during mass production is the greatest single cause of *E. coli* poisoning. U.S. recalls of tainted meat have risen to more than sixty a year.[5] Among recent ground-beef recalls: 1.1 million pounds from American Foods that had been selling on store shelves for a month in fifteen Midwestern states; 19 million pounds processed at a ConAgra plant in Greeley, Colorado, and shipped across the entire western United States; and the record—25 million pounds processed by Hudson Foods at a Nebraska plant.

In Brianna's case, even though bad meat was shipped to Sizzler, shoddy food preparation practices are blamed for making patrons sick. Once Paul Biedrzycki determined that Brianna and the others had been poisoned by watermelon contaminated by tainted meat, he had to figure out where the meat had come from. He called the CDC and the U.S. Department of Agriculture, which oversees the country's meat production, for help. Together the investigators traced labeling information on packaged meat in Sizzler's storeroom to the supplier, a slaughterhouse in Fort Morgan, Colorado. The plant was run by the Excel Corporation, the country's second-largest beef producer. The USDA investigators determined that some of the meat the Excel plant shipped to Sizzler had been contaminated with bits of cow feces, which is one of *E. coli*'s prime breeding grounds. USDA inspectors had issued twenty-six reports of fecal contamination to the plant's owners in the preceding ten months. But because food-safety rules allow processors time to respond to citations, and because inspectors wield little power to shut down facilities, the plant continued operating, slaughtering 4,000 head of cattle each day. Milwaukee Health Commissioner Seth Foldy also notes that while federal law classifies ground beef as unfit if contaminated by *E. coli* O157, whole cuts of beef are not.

Once victims of the Sizzler outbreak learned of the restaurant's unclean preparation practices, several started lawsuits against its owners for the suffering they had incurred. Connie and Doug Kriefall filed a wrongful-death lawsuit against Sizzler USA Franchise and Eschenbach and Boysa Management Company, the local firm that ran the Sizzler where they had eaten. But two years later all the cases were still tied up in court. Holding either the Sizzler or the Excel Corporation accountable was proving exceedingly difficult. Once Sizzler had been sued, it turned around and sued Excel. But in May 2002 a Milwaukee County Circuit Court judge ruled that Excel could not be sued in such cases because it had followed USDA regulations covering the handling of beef. In essence, since it hadn't actually broken any rules, Excel couldn't be sued. With such weak safety standards, few suits against meat procedures, or retail outlets like Sizzler, have been won.

Meanwhile, as companies are let off the hook, the public must live each day with the fear that this random killer might strike close to home. Connie Kriefall says she was aware of *E. coli* before, but never thought it could happen to her family. She still goes through emotional extremes over Brianna's death. Every time Chad feels bad, she feels sick to her own stomach, wondering, *Could he have* E. coli *this time?*

Brianna's death is not just tragic. It is a troubling mystery. How can it be that a common, harmless bacteria can turn so hostile and defeat our highly evolved immune system? Is there any chance that the medical research community can create a drug that beats the pathogen instead of making it more deadly? Answers to these questions—and some hope—are coming from a discovery made by a tireless microbiologist at the University of Wisconsin: Rod Welch. Welch's recent work is a triumph of modern biotechnology, making use of the latest in genetic research techniques. His breakthrough is a prime example of the kind of ingenious work we need more of if we are to win the pathogen war.[6]

Welch hypothesized that the normal *E. coli* strain must have diverged down a more pathogenic pathway, and his idea was to find exactly which stretches of genes within *E. coli* O157's entire genome made it deadly. He figured that if he could identify the mutant sections of the *E. coli* O157 genome, then drug designers could concoct a compound that would target those, and only those, novel parts. Ideally, such a drug would not unintentionally prompt the pathogen to crank out extra toxin.

Welch's notion grew out of an earlier realization by scientists that the symptoms caused by *E. coli* O157 were remarkably similar to those caused by another infectious bacteria, shigella. *E. coli* O157 and shigella victims both experienced the same onset of severe, bloody diarrhea and comparable degrees of toxic overload.

Shigella is spread primarily by individuals who have touched infected fecal matter but have not properly washed their hands afterward, such as children in day-care centers, and occasionally restaurant workers who use the bathroom. So similar are the symptoms of the two that when the Jack-in-the-Box outbreak occurred, investigators at first suspected shigella. Welch figured there must be some genetic link.

The morning after Welch learned of the Sizzler outbreak that killed Brianna Kriefall, he quickly covered the hilly walk from his house to the University of Wisconsin in Madison, arriving early and energized. This episode seemed personal because it had happened in his own backyard, and he became more determined than ever to figure out how *E. coli* functions, and how to stop it.

Welch operated from a cramped, aging office lost in a darkened back corner of one of the university's long, rectangular, yellow-brick buildings. The adjacent lab rooms where his students toiled away were typically underfunded, armored with glass beakers, test tubes, Bunsen burners, refrigerators, and a few highly sophisticated pieces of equipment. College rock music spurted from a banged-up portable radio. But Welch had one weapon in his favor: energy. In Madison, where Middle Americans with mid-length hair drive their midsized cars at moderate speeds to their mid-tempo jobs, Welch was an anomaly. He had lived for years in New York City, and the forty-one-year-old approached his work like a veteran city commuter: walk fast, talk fast, and never let up.

Welch started studying food-borne bacteria when he arrived at Madison in 1982, which happened to be the same year that other researchers first identified *E. coli* O157. Welch was drawn to analyzing bacteria because of the speed at which they multiply. To him, that meant he could conduct more experiments in any given year, giving him a greater chance to discover something new.

By the mid-1990s, microbiologist Fred Blattner, who oversees the University of Wisconsin's Genome Center in the building next to Welch's, had begun to map the gene sequence of the K12 strain of *E. coli.*[7] K12 was a harmless strain that many bacteriologists used to study how *E. coli* functioned. For six years a large team of investigators under Blattner identified and sequenced the K12 *E. coli*'s entire set of genes. Welch decided to try to sequence and compare the *E. coli* O157 genome to that of the K12 strain, in an attempt to locate the genes or mutations that had turned the bacteria deadly.

His first step was to call the CDC to get a small batch of the notorious bacteria. The government agency had sample stock from a little-known Michigan outbreak, and sent Welch a small tube of it in a nondescript package by overnight mail. He would have to grow more of the organism so he would have sufficient quantity for experiments. He transferred the bacteria onto a gelatinlike substrate in a small plastic Petri dish, and put the container in a lab incubator overnight so the pathogen would multiply. The next day he placed the freshly grown, milky white bacteria into flasks of broth for continued growth and then concentrated them into vials and stored them in a freezer in suspended animation at –70 degrees Fahrenheit. Now he had his own stock of O157.

Welch, Blattner, and a team of fifty colleagues at several institutions then approached the National Institutes of Health in Bethesda, Maryland, for money to map *E. coli* O157's genome. NIH is the major source of support for health-related research in the United States, but NIH's grant reviewers said the work sounded like a fishing expedition. Annoyed but undeterred, through other sources the group managed to put the money together, ultimately including some NIH support. They were anxious to get started because they knew a rival Japanese group wanted to sequence O157, which had caused deadly outbreaks in their country, and whichever group published the sequence first would grab the headlines—important for future funding.

Once money was in hand, Welch exhumed a frozen vial of O157 and opened it. The clump of milky white goo inside, if it were thawed and released, contained enough microbes to infect the entire city of Milwaukee. Welch clipped a tiny piece of the goo and transferred it to a flask to let it grow and recover. Now the bacteria were ready for their final transformation. He transferred them into a glass tube, and slid that into a tight chamber inside his table-top centrifuge. He then spun it at dizzying speeds. The high G forces formed the bacteria into a gelatinlike clump the size of a sesame seed at the bottom of the tube.

Welch then dipped the clump into a thimbleful of specialized chemicals that would break apart each O157 cell, leaving only the cells' DNA. He drew off a small volume of the highly concentrated DNA from the aggressive O157 organism and sealed it in a little vial.

Welch then walked his DNA from the deadly O157 sample along the grassy hilltop outside his lab to Blattner's Genome Center. Blattner's group took the DNA and for several weeks used unusual molecular techniques they had perfected with K12 to create a set of manageably sized DNA fragments, each one representing a unique part of O157's long genome.

DNA looks like a ladder with many rungs. The ladder has a famously twisting shape, like a spiral staircase. Inside Blattner's computerized machine, tiny, microrobotic devices the size of sewing pins handled the DNA fragments. This allowed another computerized instrument to read off the sequence of nucleotides, known as base-pairs, that form each rung of the ladder. Once the sequences for the various fragments had been "mapped" this way, Blattner and Welch could line up the fragments to create a complete map of O157's genes. By 1999 they had enough data to begin publishing sequences that defined significant parts of the map.

Blattner's lab computer eventually charted the string of more than 5.4 million bases that described the O157 DNA. He e-mailed the sequence to NIH's National Center of Biotechnology in Bethesda, Maryland, which maintains a database of genomes. The sequence of the harmless K12 strain he had determined earlier was there, too. Then came the fun part. Blattner and Welch sat at Blattner's computer and logged on to the NIH's genome Web page. With a single keystroke, Blattner asked the center's mainframe computer to compare the two sequences. Later that afternoon it sent him the result.

Blattner and Welch were astounded by what they saw.[8] The original K12 strain and the new, hostile O157 strain were remarkably different. They shared a similar backbone—a main string of 5,400 genes. But the more virulent strain had 177 new chunks of DNA, containing 1,387 new genes inserted at various points along the backbone. Furthermore, the virulent strain was missing 528 of the K12 genes. This was a stunning amount of divergence. How on earth had *E. coli* O157 taken on so many changes?

It was clear to Blattner and Welch, who were experienced at reading genomes, that most of the new chunks had been picked up from some other organism. They immediately began comparing these stretches to the genomes of other pathogens. Hour after hour, day after day, Welch stared at mind-numbing gene sequences on his computer screen. Sure enough, just as Welch had originally suspected, several sequences from the shigella gene were virtually identical. In all, more than 100 genes, many of them new ones, were good candidates for virulence genes—those that could contribute to O157's deadliness.

The extent to which K12 and O157 differed, and the large number of foreign genes in O157, was a revelation, as was the large number of genes whose functions were totally unknown. The findings caused scientists around the world to reevaluate some basic tenets of biology. "Many of our views of evolution are too fixed," Welch would say in discussions with colleagues. "Maybe 50 million years ago there was a single intestinal bacteria. It began to mutate, to adapt, as different species evolved—one type for birds, one type for man. Other bacteria entered the intestines, too. But they did not simply coexist and evolve along their own lines, which would be the traditional evolutionary point of view. Genetic sequencing has made it apparent that in many instances different kinds of bacteria swapped genes." Microbiologists are discovering hundreds of acquired gene sequences.

No one knew exactly how *E. coli* could have assimilated so many foreign genes. That bacteria can pick up whole genes and even stretches of genes from unrelated bacteria was a fairly new and disturbing concept. But what was even more troubling about Welch and Blattner's results was that *E. coli* seemed to value the toxin-producing genes. Searching for an evolutionary explanation for this, scientists proposed that the bloody diarrhea the toxins cause make it easier for *E. coli* O157 to spread from person to person. Another emerging theory suggested that certain peculiar viruses, called bacteriophages or simply phages, that infect bacteria could have brought the toxin genes with them.

Welch decided he could better answer these difficult questions if he could determine when *E. coli* began to diverge into the errant O157 strain. Although the O157 strain had only been identified in 1982, it might have been around much longer. In the ensuing months his team examined the two genomes, looking for buried markers that indicated points of diversion. The exercise was incredibly painstaking. Welch kept at it by reminding himself of what he regularly told his grad students: "This work can be tedious. It can be frustrating. But on any given day you have a chance to discover something new. You may not immediately realize you've found it. And you may never find it at all. But that's what you have to buy into if you want to succeed."

After months of digging, Welch and Blattner determined that the two strains had begun to diverge about 4.5 million years ago. They were amazed that in the year 2000 the strains could be so different and yet behave so similarly in most respects. The research group published its results in the January 25, 2001, issue of the British journal *Nature*, to great acclaim, just one month ahead of the Japanese team.

Welch took only a day to rest and enjoy his success. After all, his work was hardly done. He had identified the potentially offending genes, but if microbiologists in academia and at pharmaceutical companies were to devise a drug to stop the bacteria, he would have to figure out if, and how, those genes directed an O157 cell to churn out deadly volumes of toxin.

For much of 2001, Welch sat at his computer searching for clues. Meanwhile, the *E. coli* outbreaks continued, faster than ever. In October 2001, while Welch was staring at gene sequences, twenty-six people on his own campus were poisoned at a pancake tailgate party prior to a Wisconsin–Indiana football game. The Wisconsin Alumni Association held the event as an alcohol-free, ostensibly safer alternative to the usual drunken parking-lot affairs. On the menu were juice, milk, sausage, and applesauce, all of which could have harbored the bacteria.

Day after day, Welch entered the novel sequences of the O157 genes into his computer, sent them over the Web to the national genome database, then waited, hoping that someone else out there had reported a similar sequence that would tell him what the genes did. Each time he got the same return message: "No matches found." Then one day a match finally came. It was from a gene in a bacteria that causes cholera, another disease characterized by massive diarrhea. Perhaps, Welch concluded, another important part of the secret to how killer *E. coli* O157 poisons victims could be found by comparing its genome with that of cholera as well as shigella. Today Welch continues to dig into the mystery of how these gene sequences cause *E. coli* to cause disease. When the mystery is solved, there's a good chance a drug can be designed that can stop the toxins from working.

Only through work such as Welch's in recent years have microbiologists realized just how widely the pathogens around us are evolving. In 2002, Welch's team completed the genome of a third *E. coli* strain, CFT073, which causes urinary tract infections. Welch had hoped the map of this strain would help him narrow down where to look for the pathogenic sequences by ruling out large stretches of genes that the CFT073, K12, and O157 strains had in common. Not only were his hopes dashed, but his anxiety was raised. "If you had asked me a year ago how similar these three strains might be, I'd have said that maybe 5 percent of their genomes were different," he explains. "But the three strains only have 39 percent of their genes in common! This was a real surprise to the entire research community. The genetic sequences of humans and mice are more similar than these three strains are to one another. *E. coli* has acquired incredibly adaptive traits." These traits allow it to spawn offspring that are radically different. Welch says these daunting discoveries are raising a vivid warning that microbiologists know less than they thought they knew about infectious diseases.

"Are all those differences important?" he asks, frustrated. "Are they just evolutionary leftovers? We don't know. There is a whole lot more to be understood."

The type of research Welch is conducting into how particular genetic sequences cause diseases is one of the hottest and most promising areas of work in the war against deadly microbes. Indeed, a similar goal drives several hundred faculty members at universities and medical centers around the world who teach and do research into how infectious diseases operate. If investigators can discover the "virulence factors" for *E. coli* O157 and so many other deadly agents, their work may well lead to a whole new way to fight bacteria. Today's antibiotics are "broad spectrum"—they're like big bombs dropped in an attempt to kill every bacteria. The bombs may kill the enemies, but they also kill helpful bacteria in our bodies, and sometimes damage crucial cells in the gut, kidneys, liver, and bloodstream. Furthermore, our imprecise antibiotics may miss pathogens that are hiding out, or have built fortresses around themselves.

If the sequences responsible for virulence in pathogens could be determined, however, scientists expect they'll be able to design very selective drugs that will block only bacteria with the offending sequences. Like a "smart bomb," the drug would neutralize the toxins but leave our normal cells alone. The O157 would not think it was under attack, and therefore would have no reason to rev up its toxin production. In the meantime, our natural immune cells could destroy the bacteria without setting off alarms. Devising a "smart" antidote that could selectively inactivate the *E. coli* O157 toxin would be a huge advance. Several groups are working on this goal, but "it could take ten years," Welch says. "And that's frustrating. Kids are still dying."

Some researchers say an alternative might be to engineer a heroic *E. coli* that counteracts the villainous *E. coli.* Microbiologists in Adelaide, Australia, are working on inserting a gene into an O157 strain that causes it to absorb the same toxin that the rogue O157s produce. If so, a patient could swallow a pill that releases the new *E. coli.* It would multiply in the gut alongside the rogue *E. coli* and sop up the toxins there, preventing the poison from crossing into the bloodstream and collecting in the kidneys. This would buy time for the body's immune system or the other gut microbes to kill off the harmful bacteria. This practice of using one bacteria to fight another, an emerging science called probiotics, is being tried in lab mice and in human patients with less life-threatening illnesses, but thus far scientists are only working with natural bacteria like those found in yogurt.[9]

In the meantime, Synsorb Biotech Inc. in Canada is making an intermediary type of drug called Synsorb Pk that is intended to reduce the severity of symptoms from *E. coli* O157, particularly the progression to hemolytic uremic syndrome.[10] The drug stays in the gut and binds up toxins that *E. coli* produces there. This means that it must be used early, before the toxins spread to the bloodstream and organs. When researchers gave the drug to 152 children with *E. coli* O157, only 7 percent who got the drug within two days of onset advanced to HUS, compared with 17 percent of the children who received a placebo. Unfortunately, diagnosing the infection can take two days, and that assumes parents take the child to a doctor immediately after the disease is contracted. The FDA has not yet approved Synsorb Pk for general use.

For now, only our own immune systems can safely defeat *E. coli* O157, so the best we can do to help is to reduce our chances of being exposed to the killer in the first place. There are several steps you can take to lessen your exposure to *E. coli* O157, as well as other food-borne pathogens. Most important of these is to wash your hands often, particularly after going to the bathroom or changing diapers. Don't buy or use food that has passed the expiration date on the label. If a food product feels slimy, has a foul odor, or appears discolored, throw it out. Don't allow raw meat, poultry, or seafood juice to drip onto other groceries in your shopping cart. Bacteria grow quickly at room temperature and quickest at warm temperatures. Keep your refrigerator at 40 degrees Fahrenheit or below, and your freezer at zero. Freeze meat and poultry if you will not be using it within two to three days. Do not thaw frozen food on the kitchen counter, but in the refrigerator, or in the microwave if you will be cooking it immediately afterward. Rinse or peel raw produce thoroughly.

The most likely places for contamination in the kitchen are the sink, the cutting board, cutting utensils, cooking equipment, and the sponge. Clean and disinfect them all regularly, and immediately after any contact with raw meat, eggs, poultry, or seafood.

Wash your hands after preparing food. Cleanse utensils, sponges, and dishes at a high temperature in a dishwasher, or replace sponges frequently. Buy a meat thermometer and use it; cooking red ground meat to a minimum of 160 degrees throughout, and whole or ground poultry to a minimum of 180 degrees, can kill many food pathogens, including *E. coli* O157. Don't eat unpasteurized milk or cheese, and don't drink unpasteurized juices—or boil them for thirty seconds before drinking.

Despite your best efforts, you and your family will still occasionally inhale, eat, or come into contact with food-borne pathogens. So doctors recommend that you learn to identify important symptoms. When children are sick, be sure that they explain in detail any difficulties or discomforts they may be having. Kids often have diarrhea, and parents can't rush them to a doctor every time it occurs. But blood mixed in a child's stool is not normal, and in that event parents should contact a physician immediately. *E. coli* O157 doesn't always cause bloody diarrhea, however; ordinary diarrhea accompanied by severe abdominal pain, nausea, vomiting, headache, or fever above 101 degrees Fahrenheit is also a signal of a serious problem.

If you or your children do have diarrhea, be sure to drink enough to replace the lost fluids. Parents of young children can consider giving them drinks formulated with a high level of electrolytes, which further aid in overcoming fluid loss.

These and many more steps to reduce your risks are outlined on the website of Safe Tables Our Priority,[11] a nonprofit organization based in Burlington, Vermont, and founded in 1993 after the Jack-in-the-Box outbreak, by families and friends of victims of food-borne illnesses. It tries to assist victims' families, and advocates policies that prevent unnecessary illness, injury, and death from food-borne pathogens. STOP is also at the forefront of trying to spread more public information about where the dangers of food-borne pathogens lie. There really are no nationally coordinated education campaigns. A few enlightened cottage industries are beginning to advise consumers. Some orchards, like Bartlett's in Richmond, Massachusetts, now post brightly colored leaflets at their registers with a warning from the U.S. Apple Association that reads in part, "Cider was a staple of the Colonial diet. And it fits right into today's healthy lifestyles. However, the federal Centers for Disease Control and Prevention advise that concerned consumers can reduce their risk of *E. coli* O157 infection by first boiling fresh cider."

A brief description of symptoms for a dozen common food-borne illnesses can also be found on STOP's website, www.STOP-usa.org. But the burden of improving food safety belongs squarely on the shoulders of industry and government. The food industry must clean up its act. Better cleansing technology is just a start. Excel Corporation, which sold the bad beef to Sizzler, is developing a new steam pasteurization process that laboratory tests indicate could kill almost all bacteria such as *E. coli* O157 on meat. The company also recently began irradiating some of its beef to kill pathogens. Of course, these added processing steps would add to a manufacturer's cost, and whether the industry implements such measures widely, if at all, remains to be seen.

The greatest problem is weak regulation of the food industry. We must demand that our government impose strict standards and enforce them much more strenuously. Congress has met stiff business opposition to food controls. For years the FDA admitted that the nation's food regulation system needed an overhaul. It had become a patchwork of rules that had evolved over a century and was fragmented among a dozen agencies. After the September 11, 2001, attacks on the World Trade Center and the Pentagon, the FDA sounded the alert that this inadequate framework left the U.S. food supply vulnerable to terrorist attacks. Democrats and Republicans came together in the House and Senate to draft legislation that would have increased inspections of imported foods, authorized the FDA to detain food shipments without a court order, and let the FDA inspect company records to determine the source of tainted foods. But by April 2002 the legislation had become hopelessly stalled in the committee that was to work out the final details after heavy lobbying from the National Food Processors Association, the Grocery Manufacturers of America, and the Food

Marketing Institute, which claimed the legislation went beyond what was needed to confront terrorism. Meanwhile, the U.S. Department of Agriculture acted on its own to help protect the country to some extent by issuing guidelines for increased inspection and tracking of imported foods.

But a regulatory overhaul is exactly what is needed, and the safety of meat is the place to start. Unfortunately, the USDA has contradictory missions and no teeth. The agency is charged with keeping meat safe, but also with promoting the meat industry. "The marketing objective is the older of its two responsibilities, and it dominates the agency's approach. They don't want to inconvenience meat companies or cause them to lose money or sales," says Karen Taylor Mitchell, executive director of STOP.[12]

The USDA has the authority to fine and shut down plants, but only if the plant is breaking a law, and vagaries abound. It isn't against the law, for example, for a side of beef hanging in a plant to be contaminated with salmonella—a law is only broken if an employee finds it and intentionally ignores it. Furthermore, critics maintain, the laws are weak and the USDA seldom applies its powers. Repercussions are often mild: If an inspector in the USDA's Federal Meat Inspection Service finds a problem and the meat processor doesn't fix it, the agency's common action is to withdraw its inspectors from the plant, so the company's products cannot carry the "USDA Approved" seal.

Apparently the USDA even faces resistance by the courts. In December 2001 the U.S. Appeals Court for the Northern District of Texas upheld the finding in *Supreme Beef v. USDA* that the USDA overstepped its bounds when it tested ground beef at a Supreme plant for salmonella.[13] The case began when the USDA told Supreme Beef it wasn't going to pass inspection. The agency gave the company several opportunities to improve plant conditions, but, rather than meet inspection criteria, Supreme sued the USDA for overstepping its mandate. The USDA argued that high salmonella levels were indicative of a filthy environment, but Supreme denied it. The USDA has authority to withhold approval based on unsanitary conditions, and in one set of routine tests, 47 percent of the samples were contaminated with salmonella. Incredibly, the court decided that because restaurant cooks and housewives know they have to cook beef to kill pathogens, our health is not threatened by contaminated meat. That's a ridiculous argument, since the same could be said about *E. coli.*

The USDA cannot even require a recall once it has found contaminated meat on store shelves; it can only ask the producer to voluntarily issue a recall. And it can only be sure of compliance if it has shown that an *E. coli* sample from a person who has already fallen ill or died from *E. coli* O157 poisoning genetically matches a strain of O157 it has found on meat labeled from a specific plant. Knowing that this leap is difficult to make, producers often drag their feet on a suggested recall, if they respond at all. When, in May 2002, inspectors found *E. coli* O157 in ground beef processed and packaged at a ConAgra plant in Greeley, Colorado, it recommended the company recall meat processed during the same time as meat for the contaminated package. In June of that year, ConAgra issued an initial recall of 354,200 pounds of ground beef. Not until the CDC showed, three weeks later, that the same O157 strain was causing outbreaks in multiple states did ConAgra widen the recall, to 19 million pounds that had been processed and shipped at around the same time. At least forty-seven people were sickened and one died. Senator Tom Harkin (D., Iowa) called for an inquiry and investigators arrived at ConAgra on August 23, 2002. They discovered that only 12,000 pounds of the recalled meat had been recovered.[14] Only when a company is worried about liability due to illness and death, it seems, does it "volunteer" to make an aggressive recall. The USDA was also investigating why its inspectors took nearly two weeks to tell ConAgra in the first place that its meat was suspected of harboring *E. coli.* "Food safety should be at least as important as tire safety, for which there are mandatory recalls," says Mitchell at STOP.

Mitchell adds that the USDA should be required to do microbial testing on all meat leaving processing plants, not just to perform random spot checks as it does now. It also should be given greater power to fine and shut down plants. The USDA's conflict of interest between protecting the public and promoting the food industry should be abolished, or inspection powers should be given to a different regulatory body with no conflicting agenda.

Improving food safety would reduce a host of food-borne pathogens that are just as dangerous as *E. coli* O157. Salmonella can flourish in raw eggs, poultry, cheese, pasta, and shrimp, as well as beef. Listeria thrives in preprocessed meats like deli ham and hot dogs. In December 1998, the Sara Lee Corporation recalled 15 million pounds of these products after the CDC identified them as the source of a nationwide listeria outbreak. Dozens of people in the New York City metropolitan area fell ill during a September 2002 outbreak and six died. The following month is when Wampler Foods issued its recall of 27.4 million pounds of chicken and turkey due to listeria, the largest U.S. food recall ever. Yet meat plants are not required to test for listeria. Congressman Henry Waxman of California was so disgusted with the regulatory situation that in December 2002 he told the *New York Times,* "I think what we are seeing is a picture of a department [USDA] that has abdicated its responsibility to protect the public in the area of food safety."

In the past few years, campylobacter, a more recently discovered pathogen, has surpassed both salmonella and listeria as the leading cause of bacterial food poisoning. CDC statistics show that in the United States alone, food-borne pathogens and contaminants now cause 76 million illnesses, 325,000 hospitalizations, and 5,000 deaths a year. The CDC has determined that known pathogens are responsible for 14 million of the 76 million cases—which means it doesn't know what organisms caused the other 62 million illnesses.

Better still, Mitchell says, the government should create a single federal food agency to supplant the twelve agencies, including the FDA and the USDA, that have partial and overlapping jurisdictions. With such efficient operation, she says, "it would cost only pennies a pound to make ground meat safe. It should not be the responsibility of homemakers to detoxify meat before they can eat it." Citizens who are incensed by the country's needless food poisonings can get involved with STOP, or other watchdog groups like the Consumer Federation of America or Center for Science in the Public Interest. They can also contact legislators and ask them to exert leadership. "Many legislators are ignorant on this issue, and simply believe what meat industry lobbyists tell them," Mitchell says.

Mitchell's partner, Nancy Donley, the president of STOP, adds that unless the government much more rigorously inspects and punishes the food industry, more innocent victims like Brianna Kriefall will die, especially children. Donley knows the danger more deeply than anyone. On July 18, 1993, her son, Alex, died from hemolytic uremic syndrome, triggered by *E. coli* O157. "Alex was a fun-loving, healthy little six-year-old boy," Donley writes in a memorial on STOP's website. "Six short months after Alex's kindergarten graduation, I entered every parent's worst nightmare. I watched my child die a brutal death. The agonizing events occurred over a four-day period.

"In an effort to escape the continuous, racking abdominal cramping, Alex curled up into a fetal position and begged me to hold him. I stroked his face, attempting to calm him, to soothe him. I watched in horror as his life hemorrhaged away in his hospital bathroom; bowl after bowl of blood and mucus gushed from his little body. Later, I helped change blood-soaked diapers that he had to wear after he could no longer stand or walk. Alex's screams were followed by silence as the evil toxins attacked his brain, causing him to lose neurological control. His eyes crossed and he suffered tremors and delusions. He no longer knew who I was.

"I sat with my only child as the monitors registered organ failure after organ failure. His body swelled uncontrollably as his kidneys shut down. I lost count of the units of blood and platelets being intravenously fed to him. His little body had a hole dug into his side where the doctors frantically shoved a hose to reinflate his collapsed lung. Holes for brain shunts were drilled into his head to relieve the tremendous pressure. I screamed for the nurses as he suffered a massive seizure that left him on a respirator. I watched his brain waves flatten. My vibrant little boy, with his beautiful red hair and heartwarming smile, was reduced to a shell of a corpse as his father, his doctors, and I all stood helplessly by.

"Alex's last words to me were, 'Don't worry, Mommy,' as I couldn't stop the tears from silently flowing down my cheeks. His last act before slipping into a coma was to blow a kiss to his father.

"Tom and I asked to be alone with Alex after he was pronounced dead. Then, just as we did on the day he was born, we completely undressed him and memorized every detail of him. And we kissed him goodbye, our own hearts forever broken.

"From the age of three, Alex wanted to be a paramedic so that he could help people. So when he died, we wanted to donate Alex's organs, to fulfill his wish of helping others. We were told we couldn't. The toxins produced by *E. coli* O157 had destroyed all his internal organs.

"When I learned that Alex died because contaminated cattle feces lurked in the hamburger he ate, I was shocked, horrified, and incredibly angry. I felt betrayed by the meat industry, by the USDA seal of approval, and by my God. In response, I was determined to do whatever possible to ensure that others wouldn't have to go through the brutal suffering and death that Alex went through, and so other parents wouldn't have to live in constant grief and pain as Tom and I do. I learned that my goals were the same as STOP's—to prevent unnecessary illness and death from pathogens in food. We prevail on industry and government to put more preventive measures in place to keep harmful pathogens out of our food.

"It took me a while to come to terms with my sense of betrayal by God. I realized that God doesn't cause *E. coli* O157 to contaminate food. He allows free will in the world. It is ineffective government regulations and corporate greed that allow food to be contaminated."

CHAPTER SIX

BACTERIAL RESISTANCE: *The Dangers of Antibiotics and Hospitals*

Football fans might be familiar with the dramatic story of Shane Matthews, which was widely reported in the media. He was the Chicago Bears' starting quarterback for the autumn 1999 season. At six feet three inches and 204 pounds, the twenty-nine-year-old threw hard and scrambled fast. Massive, angry linebackers would grind him into the turf when they sacked him, but Number 9 would get up, shake off the hit, and call the next play. As the Bears' season ended, though, Matthews found himself nursing a hernia, which happens to professional football players. He went in for double-hernia surgery as soon as the season was over. Doctors cut open his groin, repaired the damage, and sewed him up.

Matthews was sore for the first few days. He was back home in early January 2000, but then he got sick.[1] He went to his doctor for tests and the results indicated he had a staph infection. Dangerous staph bacteria creeping around at the hospital had gotten into his surgical wound. A doctor's hands might have been contaminated with unseen bacteria from a previous patient and rubbed off onto Matthews's raw scar following surgery. Or the pathogens might have been lurking on the counter of a nurse's station, left there when a nurse who had just changed another patient's bandage rested her hand there without washing it. Another unsuspecting nurse who was preparing to change Matthews's dressing after surgery could have then placed her hand on the same counter before going into his room.

However it invaded, the bacteria quickly established a beachhead in Matthews's body. His doctor prescribed an antibiotic. It did nothing. He put Matthews on another drug, then a mixture. They did little. Matthews spent the next three months on his back in the same hospital, and nearly died. By the end of April 2000 his weight had dropped to 174, but finally his condition started to improve. It took that long for even his big, tough body to beat back the staph bacteria that the drugs didn't seem to touch. Finally back home, he could barely walk to his mailbox. It took ten months for him to recover, but he found himself a weaker and slower player. He was now a free agent, but no team wanted him. The Bears finally took him back as a third-string quarterback, with little intention of putting him on the field during a game. After that miserable season, Matthews went on a crash program to rebuild his beaten body, and two years later, in August 2002, he clawed his way back to being a starting quarterback, this time for the Washington Redskins.

Why didn't drugs defeat the bacteria infecting Matthews? Staph is nothing exotic. It's been combated for years by antibiotics. But unfortunately for Matthews, the staph that invaded his body had developed resistance to the drugs commonly prescribed to beat it.

Just like *E. coli,* staph and many other pathogens are evolving in scary ways. The more researchers investigate, the more fast-changing microbes they find. One of the most troubling features of this rapid evolution is that many bacteria are learning to resist more and more of the antibiotics we have to fight them. The uncomfortable truth is that the drugs that have so effectively countered many deadly bacteria are losing their power. Without their help, even people who have robust immune systems can be seriously sickened or killed by resilient bugs, and people with compromised immune systems face a significant chance of dying.

A bacteria can mutate any time it reproduces, and all of that bacteria's offspring will bear that mutation. Indeed, because bacteria replicate so often, random genetic mutations are common, and some of those mutations create genes that arm bacteria with drug resistance. Alternatively, bacteria can acquire already complete drug-resistance genes from other bacteria. A change in even just one gene can give a new strain of bacteria the ability to fend off a given antibiotic, maybe even all of the antibiotics in a large class of drugs. Such "resistance genes" provide bacteria with remarkable defense mechanisms. For example, one set of such genes instructs a bacteria cell to build a pump that sucks up drug molecules as they penetrate the cell, and blasts the drug right back out so it never does any harm. Such genes might have arisen initially to protect bacteria from turf wars with other toxin-producing bacteria.

Resistance can begin in a single person when bacteria are only partially challenged by medication. This can occur when a patient is given too low a dosage of a drug, or stops taking it when he feels better, rather than completing the full course of treatment. Children often spit out half of the medicine they are given because of the taste. These conditions allow a pathogen to develop ways to fend off the chemical warriors. Then its progeny mutate in a way that makes them more capable of surviving higher doses of the same drug. Successive generations confer an ever-growing ability to beat the medication, ultimately creating pathogens in a stepwise fashion that resist even strong concentrations of the drug.

As mentioned in the preceding chapter, part of the problem is that most antibiotics are "broad spectrum," meaning they attack any and all bacteria in the patient's body. When a person takes an antibiotic for a staph infection, for example, the drug molecules will kill the invaders but also destroy harmless bacteria, inflicting the equivalent of civilian casualties in a war. Meanwhile, pathogenic bacteria cells that have randomly mutated or acquired resistance genes will survive in the person's body, and will face little competition. Normally, healthy bacteria occupy most of the sites in the gut, through which harmful bacteria can squeeze into the bloodstream. They also consume a large portion of the nutrients available to bacteria. In their innocent way, they crowd out pathogenic bacteria and keep us from getting sick more often. But after antibiotics have attacked, they are scarce, leaving the resistant bacteria free to feed and multiply. The healthy bacteria will reestablish themselves, but the drug-resistant pathogens will settle among them in greater numbers. And they will have evolved to better resist the same antibiotic next time it comes around.

Just as we saw with *E. coli* O157, bacteria also develop new traits by swapping genes with one another. For example, a staph bacteria—one of the most common pathogens that often lurks in hospitals—could be handed a new resistance gene by a different kind of resistant bacteria that happens to be close by, or by special viruses that infect bacteria and can take genes with them to their next host. Otherwise harmless bacteria inhabiting our gut or skin could become reservoirs of drug-resistance genes, passing them on to visiting pathogens. Genes also pass between bacteria during conjugation. Or staph could possibly take up the remains of, say, a recently deceased strep bacteria (another common fiend), and assimilate its resistance genes into its own genome. In these cases the new generation of pathogen jumps in one leap to drug resistance.

Even in the stepwise fashion, resistance doesn't necessarily take long to evolve. Penicillin was introduced commercially in 1943, and bacteria that could resist it were discovered only three years later, in 1946.[2] Tetracycline went on the market in 1948 and was being resisted by 1953. Erythromycin and vancomycin were deployed in 1952 and 1956 respectively, and remained effective until 1988, but bacteria resistant to methicillin appeared only a year after it was introduced in 1960. Furthermore, once an organism is resistant to one drug, it is likely to resist the whole family of compounds to which that drug belongs. A staph strain that resists penicillin is also likely to resist ampicillin and amoxycillin. Tweaking the molecular structure of this family of drugs will not provide much of a counterattack, because in a few years staph will become resistant to the whole family again. An entirely new type of drug is needed to combat the pathogen.

And yet, for decades after the invention of antibiotics, neither doctors nor microbiologists were overly concerned about the dangers of resistance. The drugs made such a huge difference in treating patients in the 1940s, 1950s, and 1960s that doctors simply accepted the relatively low percentage of drug-resistant cases as inevitable. By the 1970s, pharmaceutical companies had devised so many different drugs that everyone figured a doctor could always just switch to an alternative drug if his patient's infection seemed to resist his first choice. The attitude that science and industry had such a deep arsenal that they could stay ahead of the bugs became so pervasive that the pharmaceutical industry dramatically cut back research budgets dedicated to finding new cures.

This was foolish, and not just because it weakened our supply chain of medications; the lack of emphasis on new antibiotic development also weakened our infrastructure for creating new weapons. From 1965 until 1999—thirty-four years—the FDA did not approve a single new class of antibiotics.

We now face a near-crisis situation. For any one person, battling a drug-resistant bacteria is a dire yet isolated event. But because so many people are taking the same drugs to fight their battles, the entire kingdom of bacteria is reaching a higher level of sophistication. Staph is not the only pathogen that now resists multiple antibiotics. Resistant variants of three life-threatening organisms—*Enterococcus faecalis, Mycobacterium tuberculosis,* and *Pseudomonas aeruginosa*—already defy nearly every antibiotic doctors can throw at them. "Multiple drug resistance is getting beyond individual patients or closed communities like hospitals and spreading across the general public," says Keiichi Hiramatsu, professor of bacteriology at Juntendo University in Tokyo. "People are carrying drug-resistant strains around without knowing they are doing so." In the United States alone, *E. faecalis* now causes 1.4 million illnesses and 500 deaths a year.

The pace of the war is alarming. In 1999, Pharmacia introduced a new drug named linezolid, under the brand name Zyvox.[3] It was the first new type of antibiotic to hit the market after the thirty-four-year hiatus. It killed bacteria by sabotaging internal communications inside a bacterium cell, instead of interrupting the bacterium's production of a cell wall, as penicillin and many other previous drugs had. As such, microbiologists hailed it as a promising new weapon. Staph had evolved to resist penicillin and then two of the other most commonly prescribed antibiotics, methicillin and vancomycin. Staph and ten other bacteria also resisted erythromycin, clindamycin, tetracycline, and trimethoprin-sulfamethoxazole. Because linezolid worked differently, researchers believed it could stop the evolution of pathogens resistant to all other antibiotics. In April of that year, Kathleen Fleming, a seventy-eight-year-old in England, was given linezolid on compassionate grounds while the compound was still in the final stage of human trials. Fleming had contracted salmonella, which led to a perforated bowel that let all sorts of bacteria into her bloodstream. The infections resisted all drugs; her doctors literally told her there was no other medication they could give her. Then they got the okay to try linezolid. It saved her life.

Yet a mere two years later the wonder drug was already beginning to lose its effectiveness against certain bacteria, such as enterococcus, which had developed resistance.

How did we end up in this dilemma? More and more bacteria are creating and swapping resistance genes because we are pressuring them into it. Bacteria have been on earth for billions of years, but mankind's sudden, widespread introduction of antibiotics in a few short decades has forced the creation of superbugs we may no longer be able to defeat.

The looming threat of incurable bacterial infections is an "international public-health nightmare," says Stuart Levy, a professor of medicine, molecular biology, and microbiology at Tufts University, and one of the world's best-known experts on drug resistance.[4] As infants and adults everywhere cough, burp, kiss, eat, go to the bathroom, have sex, and travel from country to country doing all these things, the drug-resistant bacteria are spread throughout local populations and then around the globe. They intermingle and share their genes for antibiotic resistance. As Levy says, "The exchange of resistance genes is so widespread that the entire bacterial world can be thought of as one huge organism that is adapting against our arsenal of drugs."

One reason we've found ourselves in this precarious situation is that the medical community was reluctant to believe that bacteria really could develop resistance so quickly and effectively. Some of the most persuasive evidence that raised the alarm was produced by Hiramatsu at Juntendo University.[5] He was concerned about drug resistance long before most scientists. His worry had been raised by an infectious bacteria called staphylococcus, common around the world. Staph infections, as doctors refer to them, exist primarily within the nose, but also appear in the underarms, the vagina, the pharynx, and on damaged skin. Staph is a leading cause of pneumonia acquired during a hospital stay, and of surgical wound infections like the one that struck Shane Matthews. By the mid-1970s, one species of staph, called *Staphylococcus aureus,* was becoming resistant in the United States and Britain to methicillin, the primary drug used to treat the bacteria. Then, in the mid-1990s, staph aureus began to become resistant to vancomycin, reserved as a drug of last resort to treat patients who had methicillin-resistant infections. But both trends had arrived forcefully in Japan ten years ahead of the rest of the world. Hiramatsu was already on the case.

Other researchers who had been studying how staph aureus's drug resistance arose had discovered that resistance to methicillin was spread not only to their offspring as a normal mutation would be, but was part of a section of genes that the bacteria could also transfer to other staph aureus. The staph bacteria, the proposal went, had actually developed a mechanism for directly passing around the resistance genes to the rest of the staph population, not just their offspring. In effect, the bacteria had figured out how to package copies of their resistance genes in a microbial cassette, and had devised a way for the cassette to leave one bacteria's genome and insert itself into a new bacteria's genome—sort of like songs that are passed around on a computer disk and downloaded into the main memory of other machines.

But now staph in Japan seemed to be learning how to resist vancomycin as well. Hiramatsu had a hunch that vancomycin resistance worked differently, because it seemed to be developing in stages. He had collected samples of staph from patients who had varying degrees of vancomycin-resistant disease; one sample showed complete resistance, but dozens more were partly resistant. Although most researchers were arguing that vancomycin resistance had arisen in the usual random way, this pattern suggested to Hiramatsu that staph had taught itself how to turn into a superbug—one that could fight off multiple drugs. Could it be that this bacteria was developing resistance mechanisms that aggressively? Hiramatsu set out to answer that question.

Hiramatsu decided to follow his inkling in 1996 after a former student of his, Satoshi Hori, who had become a respiratory disease specialist at the Juntendo University Hospital, called to ask advice on how to treat a problematic lung cancer patient. After a series of cancer treatments, the sixty-four-year-old patient had developed pneumonia caused by staph aureus. The doctor had put the man on vancomycin. For eight days the drug gradually cleared his chest of phlegm. But on days nine through eleven, the man's health deteriorated rapidly. His chest X rays became cloudy, his lungs were inflamed, and he developed fever. He was plummeting toward death. Together, Hiramatsu, Hori, and other physicians decided to give the man an unusual, potent combination of drugs, in an experimental, last-ditch effort to save him. It worked.

During the tense days, Hiramatsu asked his former student for a sputum sample from the patient. He had a hunch that the drug probably killed off the regular staph aureus pathogens as well as those that could only resist low doses of vancomycin, but over those eight days the small number that shared a more sophisticated resistance method managed to multiply and attack the man.

To test his theory, Hiramatsu would examine the staph in the man's sputum sample alongside staph aureus strains he collected from fifty different patients around Japan that were methicillin-resistant and partially resistant to vancomycin. The strains were stored in one of the freezers in his Juntendo lab. He theorized that within this group there were some rare variants that already had partial resistance but would develop full resistance if he challenged them again with vancomycin.

Hiramatsu put a dozen of his lab staff and grad students on the project. They grew staph aureus cells from each of the patient's bacteria samples in the presence of low doses of vancomycin, such that the drug would kill off the susceptible bacteria yet allow enough time for a few rare cells to strengthen their own resistance. It was a painstaking procedure. Day after day, Hiramatsu arrived on the gray elevator to his lab wearing his usual white shirt, suit jacket, and tennis sneakers, prodding his staff along. Finally, after a month of investigation they found that in each patient's sample some individual cells were indeed resistant to low concentrations of vancomycin. And they were just like the ones that had staved off drugs in the lung cancer patient and then multiplied. They all seemed to have arisen from a common ancestor that was present in Japan but not in other parts of the world.

Further analysis showed that staph had not simply picked up its high level of vancomycin resistance from another bacteria, one likely possibility. The methicillin-resistant staph had actually created vancomycin resistance stepwise starting from a particular variant that was widely distributed across Japan. It was turning itself into a superbug.

The results were so potentially controversial that Hiramatsu decided to e-mail a fellow bacteria expert, Fred Tenover, at the CDC in Atlanta. "I want to publish a paper about these findings," he told Tenover, "but I am worried no one will believe them." Tenover agreed to try to replicate the experiment. Hiramatsu sent staph samples to Atlanta, and Tenover reached the same conclusion. The two scientists submitted a paper to *The Lancet,* one of the world's leading medical journals, but the editors showed little interest. They then submitted the paper to the *Journal of Antimicrobial Chemotherapy,* which published it in July 1997.

The paper created a significant stir worldwide but received virtually no attention in Japan. The reluctance of his own countrymen to accept his evidence that staph was evolving into a superbug frustrated Hiramatsu. Japanese infectious disease officials claimed that there was no staph aureus that resisted vancomycin. Hiramatsu could not believe the officials did not see the obvious threat to the Japanese people, or how a rising tide of superbugs in Japan could also threaten people in other countries, given the ever-increasing world travel of pathogens taking a free ride on humans. To get their attention, he determined he would need to show more proof. He decided he would sequence the entire genome of two methicillin-resistant bacteria, one of which fully resisted vancomycin and one of which was sensitive to vancomycin. Comparing the two genomes could perhaps tell him how staph that had already resisted one drug went about becoming resistant to a second.

No staph aureus genomes had been sequenced anywhere in the

world at that point, and now Hiramatsu wanted to sequence two. He and a colleague assembled a team of thirty-five investigators from a dozen universities and institutes across Japan. They labored in silence for a year, then wrote up their results. This time *The Lancet* agreed to publish. Their paper hit the streets in the journal's April 21, 2001, issue, and the microbiology world cheered. The team's work was a tour de force.[6] It laid out the entire genome of methicillin-resistant staph and methicillin-sensitive staph, and compared them gene for gene. The comparison showed that the resistant bacteria had not only learned how to assimilate resistance genes from other organisms, but had actually been able to create resistance mutations on their own. The bacteria were creating unique genetic sequences that could help them evolve to resist all antibiotics. Hiramatsu's work sent a loud message: The human species is in the midst of an escalating arms race with bacteria that are increasingly learning how to defuse our weapons, and are sharing their countermeasures with one another.

Seventeen months later, in September 2002, doctors reported the first case of vancomycin-resistant staph aureus in the United States.[7] They found it in a foot ulcer of a diabetic patient who also had chronic kidney failure that required frequent dialysis. He had undergone multiple courses of antibiotics for several months, including one with vancomycin. The doctors traced the gene for vancomycin resistance to another bacteria, *E. faecalis,* growing in the same ulcer. Unlike what had happened in Japan, this time the staph had acquired the resistance gene rather than make its own. The progression to multidrug resistance seems inevitable with multiple pathways to go there.

Hiramatsu's next step is to show how staph aureus actually created its newfound drug resistance. The genomes of the two samples he sequenced were 95 percent the same, yet they still differed by 3,000 nucleotides—the base-pairs that form each rung of the DNA ladder. His team is now going through them, one by one, to find the few mutations that might have created the superbug.

If staph aureus can develop elaborate mechanisms for creating drug resistance, surely other bacteria are capable of the same feats. The inexorable rise of drug resistance portends a dire future. And we have mostly ourselves to blame. Thankfully, a movement long under way to fight the problem is finally gaining notice, due to the undying efforts of Stuart Levy at Tufts.

Levy (no relation to this book's coauthor) works from an office at Tufts Medical Center on the edge of Boston's Chinatown district. For twenty years he has been preaching about the dangers of drug resistance and how to prevent it, and until recently he received little attention. He is impatient with doctors who overprescribe antibiotics; with patients who demand them unnecessarily; with farmers who recklessly feed them to animals; with sloppy hospitals; and with antimicrobial cleansers, the companies that push them, and the public that overuses them. He is impatient because each day millions of bacteria mingle on bar tops, in back alleys, in suburban kitchens and day-care centers, and on hospital counters, distributing their antidrug weapons. The microbes are shifting the balance of power toward themselves and away from us. Every hour Levy has to spend restating what must be done to tip the scales back in our favor is another hour that we drop further behind.

There is no easy way out of the drug-resistance paradox. However, we can slow the shift tremendously by stopping the improper use of antibiotics. If you have a cold or the flu, which are caused by viruses, and your doctor gives you an antibiotic, it will do nothing to the viruses. But it will kill harmless bacteria inside your body, providing unfettered space for drug-resistant bacteria to swell their ranks. And overprescription runs rampant across the Western world.

Damning statistics are everywhere. The CDC says 50 million of the 150 million antibiotic prescriptions that doctors write in the United States each year are written for patients who have viruses. But Americans tend to expect a magic bullet to cure their every ill, so they demand that their doctors give them a pill. Many doctors do just that, rather than waste time trying to explain, argue, or defend why an antibiotic won't work. More than 80 percent of the doctors attending one seminar that Levy held in Boston admitted they wrote prescriptions for demanding patients, even though it was against their better judgment. Other physicians, especially older ones, may not appreciate how much they are contributing to a bigger problem. "Before the 1990s," Levy says with scorn, "medical faculty didn't even teach students about drug resistance." Indeed, more than a few veteran doctors still do not accept that drug resistance is a problem largely of our own making. And Levy says resistance is emerging even faster in countries like Japan, where antibiotics are sold over the counter instead of being limited to prescriptions.

Some of the trouble has to do with the inexactness of doctor exams and lab tests as well; rather than force a patient to wait several days for his lab results, a doctor may put the patient on an antibiotic "just in case" the pathogen turns out to be bacteria. Data from a study led by physicians at the University of Colorado Health Sciences Center indicate that only 5 to 17 percent of sore throats in adults are caused by bacteria, yet physicians prescribe antibiotics 73 percent of the time.[8] Each unnecessary prescription strengthens the legions of drug-resistant bacteria. In many cases, the better medicine would be no medicine—but try telling that to a man, woman, or child who feels sick and wants help.

More and more troublesome hospital cases are arising in which patients become seriously ill or die because they are infected with pathogens so resilient that drugs offer the immune system no help. Few new antibiotics will be approved in the next few years because the big drug companies pinched their research pipelines for so long. But now that drug resistance is finally being viewed as a serious problem, more work is under way. A fair share of it is taking place at smaller research firms that hope to sell lucrative manufacturing licenses to the big boys. Levy is also the chief scientific officer for one such company, Paratek, which occupies the building next to his office. Paratek is developing a drug molecule that can jam the pumps that some resistant bacteria use to expel incoming molecules of the antibiotic tetracycline. The company is also looking for a compound that will turn off the genes in *E. coli* O157 that allow it to fend off drugs trying to attack it. Paratek is in the very early stages of experimenting with another alternative Levy calls an anti-infection compound that prevents bacteria from accumulating in the lungs, so they won't reach the threshold concentration needed to start an infection.

Stories like Paratek's are still uncommon, however. Although the pharmaceutical industry is beginning to focus on new antibiotics again, drug companies have not exactly rededicated themselves to finding new drugs. In fact, in 2001 and 2002 several major companies such as Roche Pharmaceutical, Eli Lilly, and Bristol-Myers Squibb, as well as several small, innovative companies, including Millennium and Essential Therapeutics, discontinued research and development into new antibiotics.[9] Paradoxically, even though the global market for antimicrobials was more than $32 billion in 2001, if the drug companies do come up with a new class of drug, like linezolid, they can't be assured it will sell well. The current wisdom—and it is wise—is to use a new drug only as a treatment of last resort, so that it does not become widespread, giving bacteria more of a chance to develop resistance to it and undermine its effectiveness. That, of course, means it probably won't be a big seller, so few manufacturers are interested in producing such new compounds.

The new science called probiotics, in which microbiologists use good bacteria to help fight off bad bacteria, may help. Probiotics are further along in Europe and the Far East than in the United States, but the work is quickly becoming international. So far, most studies have used lactobacillus, the bacteria found in yogurt, as the "good guy." It can help crowd out gut infections and possibly help prevent staph aureus in wounds from spreading. Another attempt at turning one microbe against another is bacteriophage therapy—using tiny viruses that attack bacteria. Russian doctors have used these viruses since the 1930s as a cheaper alternative to drugs. Several companies, including Biophage Pharma in Montreal, are developing bacteriophages for oral and intestinal infections.

In the absence of new families of drugs, our best defense against drug-resistant enemies is to prevent them from spreading person to person and to limit them from fraternizing. Ironically, as Shane Matthews found out, hospitals are among the most notorious breeding grounds for drug-resistant pathogens. Just when you think you are safe, your life becomes more endangered.

In November 2001 the Minnesota Health Department told all the state's hospitals and surgery centers to suspend elective knee surgery temporarily. Within the prior two weeks, three healthy young men who had undergone routine knee surgery in different operating rooms at different state hospitals had suddenly developed severe abdominal pain and collapsing blood pressure. Within four days they had gone into septic shock and died. The menace was *Clostridium sordellii,* a nasty bacteria until then rarely found in hospitals.

CDC numbers indicate that about 5 percent of people admitted to U.S. hospitals will pick up an infection there. That's 1.8 million people a year. Nearly 20,000 of them will die as a direct result, and the infections will contribute to the deaths of another 70,000. The single greatest villain, accounting for 25 percent of cases, is drug-resistant staph aureus—the pathogen that crippled Shane Matthews. More than 70 percent of all hospital staph aureus infections resist penicillin, and half resist methicillin as well. It costs more than $29 billion annually to treat these stubborn staph cases alone.

The long-simmering problem of drug-resistant bacteria teeming in hospitals is now boiling over. More than 50 percent of all bacterial infections in hospital outpatients, and two-thirds of infections people acquire while in the hospital, are no longer affected by at least one of the antibiotics that used to clear them.

Infection and drug resistance are rampant in hospitals because of the unique ecosystem inside their walls. There is a concentration of infectious patients, a large number of highly susceptible patients whose immune systems are already stressed from other illnesses or surgery, and numerous operating tables, patient beds, nurses' counters, and instruments that are constantly bombarded with germs. Drugs are administered heavily. And doctors and nurses too often touch infectious organisms on one person and carry them to the next. The higher the density of these factors, the greater the risk; patients in intensive-care units are five to ten times more likely than other hospital patients to pick up an infection.

Hospital infections are just as problematic in Japan, Britain, and numerous other countries. More than 15,000 British deaths are linked annually to the problem. Even routine surgery presents risk. In May 2000, Jim Paton, sixty-nine, underwent an operation at Wycombe General Hospital in England to repair his perforated ulcer.[10] He died five weeks later from a massive brain hemorrhage brought on by toxic shock that resulted from a wild methicillin-resistant staph infection.

Hospital staff and administrators can take several steps to reduce the throng of germs stalking their halls. If doctors and nurses would simply wash their hands with soap between patients, the incidence of infection and drug resistance would drop significantly. Researchers in the Infection Control Program at the University of Geneva Hospitals in Switzerland published a landmark study in 2000 that showed how much difference this step can make. They observed doctors and nurses for three years, noting more than 20,000 opportunities for hand washing, and tried to educate the staff along the way about how hand washing would help. At the beginning of the study the staff washed their hands only 48 percent of the time they should have done so, but by the end of the study this rate had increased to 66 percent. As a result, infections among patients decreased from 16.9 percent to 9.9 percent, and the transmission of methicillin-resistant staph dropped by more than 50 percent.

The researchers also found that hand hygiene improved significantly among nurses but remained poor among doctors, who complained that they just did not have time to scrub between patients, or that constant hand washing chapped their hands, creating abrasions that could more readily expose their own skin to pathogens. Subsequent studies have reached the same conclusion. One alternative is to provide hand-rubs laced with disinfectants at every turn. The Geneva study and others have shown that doctors are much more likely to grab a disposable, medicated towelette while moving from patient to patient than to stop at a sink to wash, and the hand-rubs do almost as good a job of killing germs.

Since the 2000 study, several manufacturers have begun offering a product that's even better—alcohol-based rinses that ooze from a bedside dispenser in foam, gel, or lotion form. A doctor simply squeezes out a dime-sized blob and rubs his hands for fifteen seconds as he walks to the next patient. He doesn't even need to wipe his hands afterward, and the compounds contain skin moisturizers. CDC tests show the rinses kill germs as effectively as soap,[11] and late in 2002 the agency recommended that doctors and nurses use the products instead of soap except when their hands are visibly soiled. Hospitals may be reluctant to switch because the rinses cost more to install and use than soap, but the products could ultimately reduce costs by lowering the incidence of in-hospital infections, which are expensive to treat. After the Veterans Administration Medical Center in Washington, D.C., installed dispensers in all patient rooms two years ago as a test, new cases of drug-resistant staph infections dropped 21 percent, and drug-resistant enterococcus infections decreased 43 percent.

Levy also suggests that hospital administrators devise plans to vary the antibiotics given to patients with the same disease, such as staph, so that a single drug is not given so widely across a given ward that it encourages the propagation of resistant bacteria strains.

But how do hospital administrators even know resistant strains are emerging inside their walls? It is hard to tell until a subset of patients starts turning sicker instead of better, and by then the damage has already begun. This problem has plagued hospitals for decades, but a young innovator, Stephen Brossette, may have a solution. Brossette obtained a medical degree, but in graduate school he got hooked on developing mathematical algorithms for finding subtle, nascent patterns in medical data at hospitals. In 2000 he founded MedMined, a company in Birmingham, Alabama, to uncover emerging outbreaks and drug resistance within hospital wards at very early stages. By the end of 2002 his company was serving almost twenty institutions.

Several times a day, a hospital's main computer sends Brossette's computer information that updates which room each patient is in, which procedures have been performed on them, and which lab results have been generated for them. By comparing what may seem to doctors to be minor or isolated changes for each patient across the entire hospital, Brossette finds warning signs that new internal infections are rising. "Say a guy was admitted to the emergency room for heart failure, and five days into his hospital stay he spikes a fever," Brossette explains. "His white cell count has elevated, and there are shifts in his respiration, and in urine- and blood-test results. And we have similar data for other patients. We know where each person has been at several points in time each day. We then look for patterns and compare them against what is normal background for that hospital, and we can say, 'A colony of drug-resistant staph aureus is growing in these specific patients, who happened to be in the ICU four days ago. The infection hasn't broken out across the ICU yet, but if you don't treat these people now, it very well could.' " Doctors can treat the patients early, and infection-control personnel can find out what is going wrong in the ICU to spread the organism.

Brossette can also show the unintended consequences of drug treatments. "By looking back at patterns for the past three months, for example," he explains, "we can say, 'There has been a 100 percent increase in resistance to drugs A and B in bacteria X lurking in ward Y.' " Administrators can change the mix of drugs doctors are giving, to reduce the problem.

An average-sized hospital pays MedMined about $100,000 a year for its intelligence, but stopping drug resistance before it breaks out can shorten patient stays, saving the hospital money. "So far the return on investment seems to be about five to one," Brossette claims. "But the real payoff is better patient care and safety."

Physicians who overprescribe antibiotics, and hospitals that spread disease, are not the only purveyors of drug resistance; the very farmers who provide our food also contribute. Levy says more than 40 percent of the antibiotics produced in the United States are given to animals. Other estimates range much higher. A small percentage goes to treat sick livestock, but the overwhelming proportion is dumped by farmers day after day into feed for chickens, pigs, and cattle they are raising for human consumption. Rather than risk losing occasional animals, and thus profits, to infectious diseases, farmers routinely add low levels of antibiotics to their animal feed. For some reason the compounds tend to make animals grow a bit faster, too. But Levy, among a growing chorus of colleagues, says this is pure waste. The concentrations at which the animals ingest the drugs are too low to prevent any determined disease. Meanwhile, the constant trickle kills off healthy bacteria, again giving the resistant pathogens more room and nutrients to multiply.

This in turn hurts the human cause in two ways. First, consumers can contract drug-resistant diseases from chicken, pork, and hamburger that is not properly cleaned or thoroughly cooked. In an FDA study done in 2001, more than 15 percent of packages of ground beef, pork, chicken, and turkey from three supermarkets were contaminated with salmonella that was resistant to at least one antibiotic.[12] Ten percent of the meat had salmonella resistant to at least three antibiotics. The widespread cases of food poisoning from bacteria such as salmonella and *E. coli* O157 will become much more deadly as these strains become more drug-resistant. And even if they don't directly cause an infection, they can pass their resistance genes to other receptive gut organisms, like *E. coli.*

The second threat is that chemical residues from the animal antibiotics survive in food even if it is properly cleaned and cooked. That means we are ingesting a very low level of antibiotic all the time. This kills drug-sensitive bacteria in our own stomachs and intestines, allowing resistant strains to get stronger still.

Tragic human cases once again illustrate the growing problem. After nine days of severe diarrhea, a sixty-two-year-old Danish woman stumbled into Bispebjerg Hospital in Copenhagen.[13] Doctors diagnosed salmonella and immediately pumped her full of their strongest antibiotic, ciprofloxacin. It did nothing. Her infection rapidly got so bad the bacteria ate a hole in her colon, allowing all kinds of bacteria in her gut to enter her bloodstream directly. Doctors infused her with two other antibiotics used to battle salmonella, but the blood infection flooded her body with poisons. Four days after doctors realized the ciprofloxacin was not working, the woman died of toxic shock.

Another twenty-four Danes had to battle the same pathogen that summer of 1998, and ten more were hospitalized, with one additional person dying. The salmonella resisted five other antibiotics doctors tried to throw at it: ampicillin, chloramphenicol, streptomycin, sulfonamides, and tetracycline. Microbiology detective Henrik Wegener at the Danish Veterinary Lab spent the next nine months figuring out what had happened, and his finding was grave: The drug-resistant bacteria traced back to infected swine. Certain local farmers had been dosing their pigs with enrofloxacin, a close cousin of ciprofloxacin, which both belong to a class of compounds called fluoroquinolones. Wegener showed that the woman and other victims had consumed drug-resistant salmonella from contaminated pork products from a single herd. The strain that killed them was able to resist the human drugs because it had developed resistance to similar animal drugs.

Enrofloxacin is used widely by the U.S. pork industry as well as the U.S. poultry industry. Ciprofloxacin is the leading U.S. treatment for campylobacter, the single largest cause of chicken-related food poisoning in the United States—more than 2 million cases annually. If campylobacter becomes drug-resistant due to overuse of enrofloxacin in animals, we will lose our leading cure. And note that ciprofloxacin is the full name for Cipro, the drug America relied on to battle anthrax. If we lose it to drug resistance, we lose a strong weapon against anthrax bioterrorism. Finally, in 2002, after three years of study, the FDA told the drug industry that the agency will no longer approve new antibiotics for animals that can potentially create resistance to drugs in humans. That's a good step. But it still leaves untouched the tons of current antibiotics being poured into animals' diets each year.

While Stuart Levy has been campaigning against the overprescription of human antibiotics and the overuse of animal antibiotics for two decades, he now finds he must rail against a third, more recent trend that has also hastened the rise of drug-resistant bacteria: antimicrobial cleansers.[14] As the general public became more aware of the mounting dangers of infectious disease in the late 1990s, manufacturers of household products exploited their fears by blitzing supermarket and drugstore shelves with new "antimicrobial" products: liquid soap for the bathroom sink, spray cleaners for the kitchen, sponges, laundry detergent, even skin moisturizers. The hand soaps got so trendy in 2000 that it became hard to find a liquid hand soap that was not labeled "antibacterial"; one typical Massachusetts store proudly displayed twenty-four styles and brands of the stuff, and not a single one was just plain old soap.

These products are not mere folly—they are dangerous. They contain trace amounts of compounds that can hurt bacteria, but the doses are so low and mild that they affect only the most sensitive germs. So, once again, these substances clear just the weakest bacteria on your hands or sink or kitchen counter, reducing competition so the pathogenic strains with higher levels of resistance can flourish. And once again, the residues that are left behind help drug resistance accumulate as the pathogenic bacteria mutate against this low-level challenge. "At best, these products do nothing," Levy says. "At worst, they alter the microbiology of your home. The weak bacteria, which pose no threat anyway, die and the strong ones, which can cause serious illness, thrive." Levy's research shows that bacteria that become resistant to triclosan, a pervasive antimicrobial agent found in many household products, also become resistant to several antibiotics.

These problems are significant for children and adults in any household. But they are particularly threatening to babies. Levy says an infant needs to be exposed to nonpathogenic microorganisms during his first year of life so that his immune system learns how to efficiently produce different kinds of immune cells. Furthermore, if a baby's environment is too clean, his immune system may overproduce a variety known as a type-2 helper T cell, which can result in lifelong allergies or asthma. The superclean home—the goal of supposedly smart parents and the vision strongly touted in consumer advertising—actually undermines a young child's ability to develop a robust immune system.

So how do we stop the great abuses that are aiding our bacterial enemies?

Physicians are the key to reducing drug resistance. Even if a doctor doesn't want to take the time to explain to a patient who wants a cure for his sore throat why there is no drug to help, he must. It might mean the doctor sees one patient fewer a day, reducing his revenue. It might mean he has to argue with a demanding patient, losing him, too. And he could even be sued if he doesn't prescribe a drug and a patient does indeed develop a serious illness.

Patients, for their part, should listen to their doctors instead of bullying them. A wide range of ailments, including the common cold, flu, and many childhood ear infections, are caused by viruses, which antibiotics don't touch. (A virus is just a clump of proteins and DNA. It has no cell wall for a drug molecule to target, and has little internal machinery for a drug to disrupt, since it relies heavily on the machinery of the cell it infects. That is one reason why there are very few anti-viral drugs.) Studies show that doctors are much more likely to prescribe unnecessary antibiotics if they sense patients expect a prescription. If your doctor tells you, "I'm sorry, but there is no drug that can help you," don't whine. Go home, treat your symptoms as your doctor recommends, and tough it out. You will get better just as quickly as if you had the drug anyway.

That sounds reasonable, of course, until you or your child is the person who is sick. Most people want medication, for fear that a routine illness is something worse; a "cold" could be bacterial pneumonia. And it is often difficult for doctors to distinguish viral infections from bacterial ones in the early stages. Many doctors will prescribe an antibiotic even if it will do nothing. But CDC studies show that in children who have recently taken antibiotics and then contract another infection, the second illness is three to nine times more likely to be drug-resistant than if they had not had the prior medication.

Walking the fine line between keeping patients happy and reducing drug resistance due to overprescription is difficult for a doctor, especially when the patient is a child, says Bonnie Fass-Offit, a pediatrician at the Kids First practice in Haverford, Pennsylvania. She was one of the few who had been trained about the perils of overprescription when she finished her residency in 1987. But when she joined her first pediatric practice in a well-to-do neighborhood of Philadelphia, parents often arrived telling her to prescribe antibiotics, and even which drugs to choose. She immediately felt the conflict between giving patients what they wanted and sticking to her principles.

After ten years of frustrating attempts to convince patients they didn't need medication, Fass-Offit and her husband, Paul Offit, chief of infectious diseases at Children's Hospital in Philadelphia, wrote a book with colleague Louis Bell to explain the situation: *Breaking the Antibiotic Habit: A Parent's Guide to Coughs, Colds, Ear Infections, and Sore Throats.* And Fass-Offit has become pushy about presenting its arguments to confrontational patients and parents. "It does take time," she says, "time you feel you could better use examining another patient who really needs your attention. And sometimes I have to explain it over and over. For a while I steered people to the book. But now our practice has published a two-page pamphlet and we give it to every new patient."

Nonetheless, Fass-Offit estimates that almost every day she is confronted by at least one patient or parent who demands drugs. "I had one mother who was a hypochondriac; she brought in her daughter all the time. On one occasion the girl's only symptom was a low-grade fever, but the mother insisted the girl had bacterial pneumonia. I told her to just watch her daughter's symptoms. She came right back for a second visit. This time I took blood tests, which revealed nothing. A week later the woman called the manager of our practice, outraged; she had gone to another doctor who prescribed antibiotics even though there was no call for them. The woman called every one of my superiors and threatened to sue the practice and me."

One father brought in his four-year-old son, who had green mucus in his nose. The father told Fass-Offit, "Every time he has this, he gets a sinus infection. He needs an antibiotic. He won't get better without one." Fass-Offit examined the boy and found no evidence of any infection, but already knew she'd have a fight on her hands. She started to explain why drugs wouldn't help, and the father immediately got angry. So Fass-Offit tried a tactic that more physicians are turning to: a delayed prescription. She gave the man a prescription on the condition that he would wait one or two days before filling it. If the boy's symptoms eased during that time, he would not fill the prescription, but if they got worse he could get medication readily. "I knew he might just run right to the pharmacy anyway," Fass-Offit says. "But maybe he would wait, and we would avoid one incident that could further drug resistance." Recent studies show that delayed prescriptions do cut down on the number of prescriptions filled. "I know parents who don't get them filled; they've considered that if their child is on antibiotics too often, it raises the likelihood that next time their son or daughter is sick, they may face drug resistance."

Doctors are using another tactic when patients have viruses. Since antibiotics do no good, yet people feel they want something, a doctor can fill out a "virus prescription pad," available from the CDC. The pad is printed with a list of actions patients can take. The doctor checks off the appropriate items, tears the page off the pad, and hands it to the patient as a prescription.

Fortunately, brave physicians like Fass-Offit, and more media coverage about the unnecessary taking of drugs, have begun an encouraging trend. Between 1990 and 2000, the number of antibiotic prescriptions written per office visit dropped by 30 percent. Fass-Offit says general practitioners are turning the tide, but she worries that certain specialists are still heavily overprescribing. "Dermatologists use lots of oral drugs for acne," she says. "Allergists use lots of drugs, too."

Make no mistake—if you feel seriously ill, you should see your doctor immediately. Don't postpone the visit; you may end up putting your life in jeopardy, as Jeannie Brown did. Go to the doctor. But then listen to what she says.

If your doctor does prescribe an antibiotic, take all of it as directed, typically several pills a day for ten days. A great many patients stop taking an antibiotic once they feel better. But the pathogens inside your body are not all dead yet. They can multiply and make you sick again. Or they may quietly spread out, swapping some genes they have gained from surviving your incomplete assault, and come back to really hurt you, more resistant to the drug your doctor originally prescribed.

Pharmaceutical companies can help limit the rise of resistance by devising new drugs more often. Levy says one big reason no new family of antibiotics was introduced for thirty-four years was that the drug companies began to look for "blockbuster" products. Drug research is costly. The regulatory approval process takes years. So, rather than invest in several dozen lines of drug research, the big companies concentrated their resources into a few products that had the potential for huge payoffs. The result is that a steady stream of specialized drugs for particular infections has been rejected in favor of a few blockbuster products that can fight a range of bacteria. This forces doctors to widely prescribe the same few medications, and it is far easier for bacteria to develop resistance to only a few broad-spectrum drugs. Vancomycin was used sparingly as early as the 1950s, and resistance was never a problem. But when it suddenly was prescribed en masse as the best cure for penicillin- and methicillin-resistant staph, resistance to vancomycin began to shoot up, too.

Interesting studies show that eliminating drugs from animal feed can lessen the incidence of drug-resistant bacteria in people who consume meat. In 1995, Denmark prohibited the use of avoparcin in chicken and pigs. Studies that year and five years later showed that drug-resistant strains of enterococcal bacteria in the country's citizens fell from 73 percent to 5 percent. In a study by Denis Corpet at the National Institute for Agricultural Research in Toulouse, France, a group of people who ate regular diets switched to eating only foods that were bacteria-free. The number of drug-resistant bacteria in their feces decreased by a thousandfold.

Animal antibiotics should be used only to treat sick animals. But it is unlikely that farmers and herders will voluntarily give them up in feed. So the government agencies regulating them must step in to ban them, and the entire class of fluoroquinolones would be a prudent place to start. In return, industry regulators could help farmers find alternative farming methods that encourage healthy animal growth, first among them cleaning up their grounds. Countries across Europe have recognized the problem, and have banned the use of animal antibiotics that have human drug counterparts as routine supplements in feed. It is outrageous that the FDA has not done the same in the United States. In summer 2002 the agency said it would no longer approve new drugs, but also said it would only continue to study the problem regarding drugs already approved. That's probably just fine with the nation's powerful meat lobbies. Indeed, Levy says, "Scientists agree this application of drugs should be banned, but they don't have the money to wield clout in Washington." In 1994 the CDC strongly advised the FDA not to approve the use of quinolones in chicken feed in the first place, but it did. Since then the incidence of quinolone-resistant campylobacter in people has risen from 1 percent to 17 percent.

Drug resistance has grown from a personal problem for a patient to an international public-health threat. Curtailing the overuse of antibiotics worldwide will require cooperation among countries. The first step is to track the emergence of resistant strains, almost the way CDC and other national labs track the emergence of rising flu strains. For two decades Levy has run a nonprofit organization called the Alliance for the Prudent Use of Antibiotics, which attempts to do just that. But in the past three years the organization has suddenly attracted wide attention, as drug resistance has become an endemic problem. The APUA has quickly expanded into an international organization to do research and help educate the public and public institutions on how to promote the appropriate use of antibiotics.

Fixing the antimicrobial-product problem is easy in theory: Consumers should stop buying them. You can limit your exposure to bacteria with a few simple steps that will not encourage the growth of pathogens or drug-resistant strains in your daily environs. Wash your hands with soap and water after going to the bathroom and before and after preparing or eating food. Wash more often if you, a family member, or a close contact has an infection. The physical action of running water literally knocks germs off your hands and into the drain, and regular soap can burst the cell walls of the rest, killing them.

Vigorously rinse fruits and vegetables to blast bacteria off them, too. Thaw frozen foods in the refrigerator, not on the countertop, and do not leave raw fish or meats or cooked foods unrefrigerated for more than an hour. Bacteria multiply faster at room temperature than at cold temperatures, and all these common practices give it more opportunity to do so. Wash all cutting boards and other surfaces and utensils with hot, soapy water after preparing foods. If you feel the need to disinfect, use bleach, alcohol, or hydrogen peroxide. These caustic chemicals kill almost all living cells.

If overuse of antibiotics in people, antibiotic feed in animals, and antimicrobial cleansers in our lives promote the growth of more-resistant pathogens and weaken the effectiveness of our drugs, then why don't we stop using antibiotics in general and just let our immune systems battle bacteria on their own? Because too many of them can beat us if we don't get help. The question is one of balance. We must use just the right quantities of the right kinds of drugs, and no more. And we must let our immune systems be challenged in mild ways so they mature and stay strong to fight the heavy battles that will inevitably arise. Our quest for a magic bullet every time we sneeze, for food industries that never lose an animal, for a sparkling, germ-free household may all contain the seeds of the next human health nightmare: a world in which drugs no longer can defeat bacteria. That would bring us back to the 1910s and 1920s, when simple germs killed thousands upon thousands of people in every country on earth every year.

In the bacterial world it is survival of the fittest, and we are killing off the less threatening ones in numerous ways that help the more threatening get stronger. Do we have the sense to forgo an unneeded antibiotic when we feel sick? To pay a little more for meat raised free of drugs? To not reach for that bottle of wonderful-sounding antimicrobial soap in the grocery store?

CHAPTER SEVEN

THE NEXT FLU PANDEMIC:
Rapidly Mutating Viruses

The alarming evolution of deadlier and more drug-resistant bacteria presents a grave challenge. That threat is joined by the escalating dangers of viruses. Just as with bacteria, killer viruses lurk everywhere, and they are even more cunning. The threat of future pandemics is brought home by the story of the insidious influenza virus. This master makeover artist mutates every year, killing 36,000 people annually in the United States and more than 100,000 people in North America, Europe, and Asia—despite widespread flu vaccinations.[1] Even more daunting, the flu virus can mutate at random into strains that can trigger runaway global pandemics.

Three times in the past century, sudden, extreme changes in the flu virus's genes have caught scientists and health officials by surprise. The 1918 "Spanish flu" killed 500,000 people in the United States and 20 million worldwide, and depressed world population growth for ten years. The 1957 "Asian flu" killed 70,000 Americans, and the 1968 "Hong Kong flu" killed 34,000. Hong Kong strains, still not fully understood, have since caused more than 400,000 American deaths, 90 percent of them in people age sixty-five or older. Ancient history? Hardly. Scientists now say we are overdue for another pandemic—and it is unlikely that any vaccine could be produced fast enough to stop it.

To appreciate how vulnerable we are to this threat, and how swiftly the world could become ravaged, consider the fate of a three-year-old boy who was admitted to Queen Elizabeth Hospital in Hong Kong in early May 1997. Twelve days later, on May 21, he was dead from an incredible combination of conditions: influenza pneumonia, acute respiratory distress, Reye's syndrome, and multi-organ failure. He had suffered a massive viral assault. The story of what happened next, pieced together in part by Jill Lee at the U.S. Agricultural Research Service, is a wake-up call that alerts us about how vigilant we must be if we are to stop that next pandemic in its tracks when it comes.[2]

Physicians were baffled over what had killed the boy, but they knew it was strange. The day he died, his primary doctor sent samples of the child's respiratory secretions to the Hong Kong Department of Health for an investigation. Lab technicians there isolated a flu virus and discovered it was a highly unusual strain. Worried that they might be dealing with an extreme genetic mutation rather than a more common, milder annual alteration, they sent samples of the strain to three leading flu labs: the CDC in Atlanta, the National Institute for Medical Research in London, and the National Influenza Center in Rotterdam, the Netherlands.

By early August the Rotterdam center had determined that the virus was of a type known as H5N1—a bird virus that had never been found in a human before. Flu experts at all the centers began to sweat. When a virus leaps from an animal species to a human, which is rare, researchers fear some deadly mutation has occurred to allow the transfer.

Humans are not the only animals that get the flu. Chickens, pigs, ducks, horses, and marine mammals can also be infected. Although a chicken strain, for example, can occasionally infect another animal, say a pig or a person, that infection usually doesn't cause illness in the new host because viruses tend to be specialized to the species they regularly infect. Usually, however, doesn't mean never. If a human shows evidence that he has indeed been infected with an animal strain, epidemiologists will immediately worry that the pathogen is a bizarre combination of animal and human virus. That can happen if, for instance, a pig becomes infected with a human strain at the same time it is infected with a pig strain. The two strains can swap genes inside the pig's cells, spawning an alien offspring that has traits of both—and therefore can infect people and attack them.

All the new strain needs to do is find its way into a human host, which isn't that difficult. Pigs sneeze and chickens cough, sending virus particles into the air, which farmers or poultry-plant workers may then inhale. More likely, the virus might be excreted in an animal's feces. A person walking past a chicken in a cage at a live-bird market, for example, might inhale bits of dried feces, with the virus riding inside. Live-bird markets crowded with customers are common in densely populated Asia, a hot spot where new flu strains arise. If an animal virus recombines with a human one, the resulting new strain can be especially virulent and set off a global epidemic, because the human immune system will have never seen anything like it, and will have few clues about how to fight it.

That's how the great pandemics have started. Scientists have carefully analyzed the genes in bird, pig, and human virus samples saved from 1918 and found that the 1918 pandemic flu was a bird virus that infected humans, and then recombined with a pig virus that had also infected humans. The mutant virus retained the traits from the human flu that allowed it to be passed from person to person, yet was still so foreign that the human immune system did not know how to fight it. Experts figure that the virus probably originated in Europe and was spread by troops returning to the United States and other countries from World War I. The virus then figured out how to move quickly from person to person, infecting millions of Americans in a short time.

The experts analyzing the H5N1 samples from the Hong Kong boy knew this history and feared they might have a similar case on their hands. They guessed that the virus had infected the boy at one of Hong Kong's poultry markets, and had combined with a human strain. They also knew that the bird virus from whence this mutant evolved was extremely deadly in poultry.

Nancy Cox was at a ranch in Wyoming in early August when her office managed to get word to her.[3] Born in rural Iowa, Cox had worked at the CDC for twenty-one years as a molecular virologist, and had become chief of the CDC's influenza branch, as well as one of the World Health Organization's top influenza consultants. She had mastered every scientific nuance and every lab procedure connected to influenza, and could patiently explain all those details with remarkable simplicity.

The ranch purposely didn't allow phones in guests' rooms, but the CDC staffers convinced the ranch owner that tracking down Cox was a matter of international importance. When Cox called back from a pay phone and heard the news, her mind went into hyperdrive. Knowing that H5N1 was a bird virus, she was alarmed that it had made the leap to humans, or at least to one human. The scary question burned in her mind: Was it contagious from person to person? Unable to get off the prairie immediately, Cox ordered her staff to stop everything and begin analyzing the virus as fast as possible. She also dispatched CDC experts to Hong Kong to check one other angle: it was possible the H5N1 variant could have arisen in the Hong Kong lab itself if technicians there had accidentally contaminated their specimen of the boy's virus with another flu strain. The CDC researchers could help Hong Kong's scientists analyze whether or not that had occurred.

Two days later Cox's crew confirmed Rotterdam's result: The virus really was H5N1. Now back at her CDC post in Atlanta, Cox had to identify the genetic mutations that had allowed the virus to make the leap to humans in order to understand the nature of the threat this virus posed and how to develop a vaccine, and she had to do it fast.

Handling a pathogen such as this one, however, requires a special lab—a biosafety level-3 lab designed to contain deadly viruses. Cox was worried the virus could be highly contagious because the bird form spread like wildfire among chickens. The CDC's influenza branch did not have a BSL-3 laboratory available, because normal human influenza viruses are handled at a lower level of biocontainment, BSL-2. Cox telephoned David Swayne in Athens, Georgia, about an hour east of Atlanta. Swayne, a veterinarian, heads the Agricultural Research Services (ARS) Southeast Poultry Research Laboratory (SEPRL) there, and led a team studying poultry influenza viruses, including the H5 types. In 1994 and 1995 the

Athens researchers had cracked the case of a vicious poultry outbreak in Mexico of the strain H5N2.

Because Swayne and his partner, microbiologist Mike Perdue, had been working with several companies to devise poultry vaccines for H5-type viruses, their laboratory had both a BSL-3 facility and sophisticated means of identifying viruses. Swayne, who readily recognized the grave threat H5N1 posed to humans—as well as to poultry, his first responsibility—immediately offered Cox his lab. By the end of August the CDC and ARS team began a collaboration that also included ARS veterinarian David Suarez. At the recommendation of CDC doctors, everyone on the project began taking rimantadine, an antiviral drug, even though when they were in the containment rooms they wore hooded bodysuits and breathed through respirators. Meanwhile, Hong Kong investigators had sent word that four other local people exposed to the virus had also died. The clock was ticking, and although only a handful of people had contracted the strain thus far, the researchers had no way of knowing whether a major outbreak might still be triggered.

Perdue and Suarez sequenced the genes of the H5N1 chicken virus from Hong Kong—a slow, methodical process that takes days—and, with the CDC, compared it against the maps painstakingly made of the H5N1 virus from the boy. Long stretches of the genes matched. Sure enough, the virus had made the species leap.

By now Hong Kong's political leaders had become increasingly worried not just about human health but about the country's poultry industry. There had been two avian flu outbreaks earlier that year, in March and May, that had killed many birds. A recent study had also shown that 20 percent of chickens and ducks at farms around Hong Kong and at the Hong Kong markets carried some type of H5N1 virus. It wasn't clear that the virus killed every bird it infected, but birds were certainly dying. A major outbreak—in chickens or humans—could not be risked. Swayne, Perdue, and Suarez decided to test several vaccines developed for other H5 poultry influenza viruses. If they found one that worked, the crisis would be averted. After injecting several test flocks, they found a vaccine that stopped the pathogen. But in early December poultry inspectors found a large number of dead chickens in a market at Cheung Sha Wan in Kowloon. Then came more news: H5N1 had claimed the life of a fifty-four-year-old man in Hong Kong. Hong Kong officials had little confidence the vaccine could cure chickens that were already infected, and it would be too hard to blanket the country with inoculations. The stakes were getting too high. On December 29, 1997, they ordered a mass slaughter of 1.5 million chickens and an industrywide cleanup of poultry markets. With this dramatic, decisive action, they averted the threat of a major human outbreak.

Back at the CDC, Nancy Cox could breathe a sigh of relief. Scientists at her lab and others determined that this strain was not, after all, contagious in humans. Epidemiologists also showed that the six people who had died all had had direct contact with birds. But Cox continued to investigate the strain to find out what kind of mutation had created it, because the alteration was responsible for the virus's ability to survive in people. Several more years of research at Cox's and Swayne's labs, and in Hong Kong and other locations, was required to figure out how the virus had arisen. First of all, they determined that the pathogen was not a bird virus that integrated some genes from a human virus, as they had thought. This strain was entirely of chicken origin. One small mutation allowed the chicken virus to infect and destroy human cells. The aggressive pathogen didn't even need any human virus genes! Cox and her colleagues were impressed that such a simple change could allow an animal pathogen to leap to humans.

Although six people died, the new virus probably did not spread from person to person, because it had not yet perfected a way to do so. Ironically, researchers surmised, because the bird strain killed its human hosts so quickly, it was not in a person's body long enough to recombine by chance with a human flu virus and mutate into a strain that did indeed have traits from both. To do that, the bird virus and the human flu virus would both have to infect the same cell within a person. That, if it had occurred, could have made the killer contagious, and started a pandemic.

The danger from this virus is not over. In June 2001 and again in May 2002 Hong Kong officials ordered slaughters of millions more chickens when the virus resurfaced. Experts think the virus was so widely spread the first time that eliminating it will be difficult if not impossible. By the end of 2002 the CDC had devised three human vaccines against the strain that were ready for testing. Nancy Cox called for a stockpile of such vaccines as they were approved, just in case the virus did start spreading in humans.

The Hong Kong government's decisive action to slaughter more than a million chickens may well have prevented what could have been the twentieth century's fourth pandemic, by denying H5N1 time to spread and evolve into a contagious state in humans. In this case both the poultry industry and the public were at risk, so the poultry industry was willing to sacrifice the flocks that might have been harboring the virus. But what if, next time, a similar scenario slowly reveals itself in the United States? Would the powerful U.S. poultry industry quickly agree to destroy its flocks? The public would like to think the industry would act for the common good, but that is not always the guiding principal of big business. Even foot-dragging to delay the slaughter could allow a deadly virus to jump to many more people than this one did, giving it more time to mutate, combine with a human strain, become contagious, and spark a global disaster.

Could another flu pandemic really be lurking just around the corner? "It is very likely," Cox says. "We just can't tell when. It could be fifteen years from now. It could be one year from now."[4]

The best way to stop a pandemic before it spreads, she says, is to test any and all flu strains that appear even the slightest bit unusual, and that is exactly what her lab at the CDC does. Without sufficient advance warning from such global monitoring, the CDC and vaccine manufacturers would not have enough time to assess the new virus properly and devise an effective, safe vaccine.

Cox delivers this warning in stern language, as a lesson borne from a debacle a quarter of a century ago. In February 1976 an army recruit at Fort Dix, New Jersey, died mysteriously from an influenza virus similar to a virus that was causing illness in swine. Testing showed that another half dozen of the post's 500 troops had it. The CDC hastily concluded that the virus represented a major genetic shift. The American population had no immunity to the virus, and officials feared it could cause a national pandemic. The CDC proposed a $135 million crash program to inoculate 210 million Americans. On March 24, President Gerald Ford addressed the nation and said "every man, woman, and child" should be vaccinated. The CDC had to devise the vaccine formula rapidly, and just as quickly the pharmaceutical industry had to prepare it.

The inoculations commenced on October 1, after a prolonged delay to work out liability coverage for the manufacturers. But after only three weeks, several dozen people had died of complications possibly linked to the vaccine. The program was abruptly halted, but the vaccine had hospitalized and sickened thousands more. Many victims developed Guillain-Barré syndrome, which is a progressive paralysis. In most people the damage is temporary, but in some there is a permanent nerve or muscle damage. More than 5,000 victims sued the government, with claims reaching $1.7 billion. And the expected swine flu epidemic never materialized. Only thirteen mild flu cases developed at Fort Dix, and the strain was not found outside the base.

Various investigations concluded that the problems caused by the swine flu vaccine occurred because the vaccine was produced in too much of a rush. Tests were not yet conclusive about the right dosage. Side effects were not thoroughly explored. And the CDC's alarms that the Fort Dix virus represented a dangerous genetic makeover, and that it was highly contagious, were not well founded. The CDC has learned its lessons, which is why Cox's group is so active year-round. It is not a question of whether the next killer flu pandemic will occur, but when. We are overdue.

Nancy Cox and her team at the CDC run a remarkable program that tries to predict which strain of flu will strike, and instructs manufacturers about the correct vaccine to develop. Every time the flu changes, scientists must devise a new vaccine formula.

The annual flu changes frequently originate in Asia, because so many people in some parts of Asia live in such close quarters and healthy living conditions may not be as uniform, raising the likelihood that a new strain will begin to spread. They then circulate from person to person around the world in a somewhat predictable manner. All year long, local hospitals and labs in numerous countries test throat swabs and blood samples from ill patients to see which flu strains people are carrying. If a strain matches a known type, the lab reports it to a state or regional health department. If the strain does not match, the lab sends the sample to the larger authority for further analysis. If that lab cannot identify the strain, it sends the sample to one of four powerful institutions around the world for exhaustive, sophisticated testing. In the United States, that institution is the CDC.

The Centers for Disease Control and Prevention employ 8,500 people around the country, 4,000 of them at six campuses in Atlanta. Headquarters is on Clifton Road in the city's northeastern suburb. It is an unlikely place for the world's most concentrated site of infectious disease organisms and experts. The circular twenty-eight-acre campus is built into a mildly sloping, lightly wooded hillside, surrounded by a two-lane street typical of any suburb in any eastern U.S. city, sporting strip malls, pizza parlors, convenience stores, a public school, and a hospital. From the shrubbery-lined sidewalk the low-rise concrete, glass, and brick CDC buildings look typically commercial. But step onto the campus, past the concrete barriers protecting the front door, and you are met by armed guards. If you don't present the right ID, you are sternly questioned, if not bluntly turned away. If allowed in, you pass through an X-ray metal detector, and so do your bags. Then a guard stands by your side until a staff member arrives to take responsibility for you.

Although the campus looks benign from the perimeter, its organization is militaristic. The seventeen units are arranged in a concentric pattern. Freestanding, open-air catwalks three and four stories off the ground are the only connections between buildings, and link the backs of buildings so the walkways are not visible from the exterior road. A guard with a pistol waits at each end of each catwalk, and any employee who approaches one of them without the proper ID tag hanging from his neck is stopped. Cameras in the ceilings monitor all passersby.

All this security is necessary because inside these labs, the world's worst pathogens are grown, examined, dissected, and tested. And no one will say where, inside, the stuff is actually stored. The CDC maintains vast stores of invasive Group A strep, *E. coli* O157, influenza, measles, polio, Lyme disease, West Nile virus, anthrax, smallpox, hepatitis C, tuberculosis, herpes, HIV, and every other germ known to man. Labs and national flu centers from around the world ship their worst influenza strains—the ones they can't identify—to Building 17, the newest on campus. The addressee: Nancy Cox.

Cox lives less than two miles from the CDC and usually arrives by 8:00 A.M. She goes straight to her computer. Overnight, leaders of labs in various countries have sent her reports listing the strains of flu they have found from the latest patient samples. There might be four cases from the state lab in Connecticut, fifteen from Spain, forty from Prince of Wales University in Hong Kong. Cox strides through the samples' telltale codes of letters and numbers, such as H5N1. On certain mornings she might also find reports from the three other global influenza centers—in the United Kingdom, Japan, and Australia—that, with the CDC, spearhead the world's battle against the disease. By assessing the 10,000 to 15,000 strains reported throughout the year, Cox can track which of the flu's many variations are appearing in which countries, and the rate and direction at which they are migrating across the globe.

By late morning Cox has reviewed unusual reports with various of her thirty-four full-time experts and thirty visiting scientists and graduate students. But she also often gets anxious. Federal Express will arrive soon, delivering the baffling influenza strains from labs that examined a patient's throat or lung culture, but could not fully identify the strain. FedEx brings mystery organisms almost every morning during the northern hemisphere's slow spring and summer flu seasons, and twice a day as new strains speed off from Asia once cold November sets in.

Cox's seven lab specialists carefully unpack the rogue virus particles, isolate them in lab equipment, multiply them to get more quantity needed for testing, and probe them to see if they behave like any known strains. If not, they are the ones that could potentially trigger a massive epidemic.

When Cox and her team determine that a new flu strain has

emerged, they then must decide what vaccine to develop to combat it. This requires a highly specialized knowledge of how viruses work. They are truly diabolical organisms. Viruses are simply tiny packets of DNA or RNA, thousands of times smaller than bacteria. They are not even complete cells, so they can't replicate on their own. And the way they reproduce—which they do at an astonishing rate—sounds like something from a Stephen King novel, or the movie *Alien.* First a virus latches onto a healthy cell inside your body. Then it tunnels into the cell and hijacks the cell's internal machinery to help it create many new virus particles. It releases its DNA or RNA into the cell, and persuades the cell to make copies. The copies then reassemble into new virus particles, which burst out through the cell's wall and burrow into more cells, spreading out and multiplying voraciously, damaging more and more cells throughout your body as larger and larger waves of offspring repeat the process over and over.

Because the invaders hide inside your cells, it is difficult for drugs to attack them without also attacking you. This is why pharmaceutical companies have devised only a few moderately effective drugs that can stop only a few kinds of viruses. The best doctors can do, typically, is try to minimize your symptoms of fever, nausea, and inflammation, buying time for your immune system to orchestrate a response.

Like bacteria, most viruses enter our bodies in predictable ways. We breathe them into our noses or mouths, we pick them up on our hands or lips or while kissing or having sex with people already carrying the germs. They stick to membranes in our nose, throat, or lungs, attach to the linings of our stomach, intestines, or genitals, and occasionally enter our bloodstream directly through a scrape or cut.

Fortunately, our immune cells are in these places, ready to act. The way our immune system works is one of the true wonders of biology. It is an elaborate, multifaceted defense organization comprising two parts: the innate immune system, and the "specific" or "acquired" immune system. The innate system is particularly effective against bacteria, and the specific system is honed to fight viruses. Understanding how these systems work helps greatly in appreciating how stealthy a foe viruses are, and how vaccines attempt to stop them.

The foot soldiers that fight the immune system's battles against pathogens are white blood cells. Often these warriors stop infectious organisms so swiftly we're never even aware we are infected. Our bone marrow constantly produces white blood cells, and it steps up production when a pathogen has infected us. From the bone marrow, most white blood cells enter the bloodstream, where they differentiate into specialists that perform distinct functions. Some specialists, referred to generally as phagocytes, come in several varieties: granulocytes, macrophages, and dendritic cells. The granulocytes are concentrated in our bloodstream and are the most efficient at chewing up invaders. The macrophages are more widely spread throughout our body; in addition to chewing up microbes, they release chemical messengers called cytokines and interleukins that recruit and activate more troops and help coordinate attacks. Dendritic cells hole up in our skin, gut, and lungs, acting as sentinels that alert the rest of the body to bacterial and viral invaders. Despite the specialization, the two immune systems and the three types of phagocytes all work beautifully together to control a given infection.

Phagocytes look a bit like microscopic marshmallows—some spherical, some irregular, some star-shaped. Their surface is irregular with multiple mittlike extensions. They ooze steadily throughout our bodies, brushing up against passersby, using their extensions to check for surface textures that they have learned through evolution might indicate foreign cells. Bacteria are mostly spherical or rod-shaped, and are about one-hundredth the size of a phagocyte. Viruses are ten to 1,000 times smaller still, and look like tiny globes, soccer balls, or even snakes. Both types of pathogens wear thin coats that contain some unique proteins.

When a phagocyte senses what seems to be a bacteria or virus, its mitts snag the invader's coat and investigate the proteins in it to determine what kind of organism it might be. For example, many bacteria, like *E. coli* O157, construct their coats with carbohydrates that human cells do not use—an immediate tip-off.

Instantly the phagocyte pulls the invader into itself in a bear hug, wrapping its pliable body around the foreign organism. Before the opponent can break free, the phagocyte imprisons it in a sac inside its own body. The phagocyte then assaults it without mercy, pouring enzymes into the sac that break the pathogen into molecular bits, or squirting in bleachlike compounds such as hydrogen peroxide, superoxide, and nitric oxide that dissolve the foe. A phagocyte can dispose of a pathogen in this way in a matter of minutes.

Phagocytes are very good at identifying and killing bacteria. But they have a more difficult time determining without doubt that a virus is an invader. Because viruses quickly burrow into healthy cells, the guards may not even see them. Most often, the confrontation begins after a virus has already infiltrated a cell. As the virus's offspring burst out through the cell's wall, the mayhem draws the attention of phagocytes, which then grab hold of the nearby suspects and digest them as they would bacteria.

However, the phagocytes know they have probably already missed escaping viruses. So the macrophages or dendritic cells preserve some of the virus bits, called antigens, and head for the immune system's closest base camp, a nearby lymph node. As they depart, they send out cytokine or interleukin messengers that speed ahead to the base camps to alert a special part of the specific immune system: the lymphocytes.

Lymphocytes are another variety of white blood cell. Instead of fighting, they operate more like CIA or FBI agents, gathering and analyzing intelligence data that they can use to tell warriors where and when to strike. Some lymphocytes patrol mucus membranes along the respiratory, urogenital, and digestive tracts, while others wait in base camps at lymph nodes spread throughout the body, in places such as the tonsils, and the neck beneath the jawbone. The lymphocytes ride from camp to camp through the lymphatic system, a series of interconnected waterways that percolate like underground springs through our tissue and converge at the camps. Other lymphocytes convene in the spleen and patrol the blood vessels.

When a macrophage or dendritic cell arrives at a base camp, it brushes up against the lymphocyte investigators, known as B cells and helper T cells, and displays the antigens. There are thousands of varieties of B and helper T cells, and each variety recognizes only one kind of antigen. The cells check out the evidence. It's something like checking a suspect's fingerprints against a database. When the group has produced enough information to determine the kind of virus it has, the B cells that are programmed to fight the invader instruct their genes to produce a weapon for the upcoming confrontation. The helper T cells also instruct appropriate "killer" T cells to ready themselves for battle.

The B-cell weapon is a protein known as an antibody. Each B cell that has recognized the antigen churns out thousands of antibodies. These molecules flow through the bloodstream and collect at the site of infection—for the flu, the membranes in your nose, sinuses, and lungs. The antibodies bump up against various cells until they feel the correctly shaped antigens, indicating the virus particles. The antibodies lock on to the antigens, disabling the virus particles one by one. With the right antibodies attached, the virus cannot invade another cell, and therefore cannot reproduce. Now the macrophages and granulocytes, which had refrained from attacking earlier, swoop in. The fighters find the locked-on antibodies and engulf the virus particles they are attached to, chewing them up or blasting them with peroxides or bleachlike compounds. Killer T cells also destroy infected cells that have become virus factories, crucial to controlling the infection.

As the battle rages, cytokines act as beacons, recruiting more phagocytes and signaling more B and T cells to join the fight. They also instruct the bone marrow to produce more white blood cells. The count of white cells rises in the bloodstream, one telltale sign that the immune system is fighting an infection.

The cytokines also contact cells outside the immune system to help. The unpleasant symptoms you suffer during a bout against flu are by-products of their work. When germs first land in your nose or lungs and are interrogated by phagocytes, the cytokines they release trigger the nose to sneeze and the lungs to cough, in attempts to literally jettison the invaders from your body. Other molecules tell nearby blood vessel cells to relax their normally tight containment of the blood so the warriors floating by can easily exit the bloodstream and flow into the infection site. This inadvertently allows small amounts of red blood cells and plasma, the fluid of the blood, to leak into the surrounding tissue, which leads to the redness and swelling of infected areas. Some cytokines also reach the brain, which responds by producing fever. The immune system works more efficiently at a slightly higher temperature—although the fever signal can overshoot in severe infections, sending body temperature too high, which can stress organs or the brain. The cytokine messengers also cause nasal membranes to produce mucus to try to flush out reproducing viruses physically, giving you a runny nose. Other cytokines induce muscles to ache and make you sleepy, forcing your body to reduce movement so it can concentrate all its energy on battling the enemy.

And so the war is waged. If all goes well, you win. Antibodies that didn't find a target will linger for months or years in the bloodstream, and in the linings of the respiratory or intestinal tract, waiting to tackle more enemies if they appear.

It is impressive that your immune system can respond so deliberately, given the many types of pathogens that can invade your body. B and T cells must analyze each intruder to determine its identity, then figure out exactly how to stop it from reproducing and how to kill it, without killing you. There is an incredible range of possibilities. The challenge is so daunting, however, that the system often needs a week or more to analyze new antigens and mobilize troops. In the interim, you suffer the effects of a cold or flu; the pathogens multiply fiercely or produce toxins, as *E. coli* O157 bacteria do. The phagocytes do what they can to slow the invasion, but the immune system begins to reverse the tide only when the B and T cells have reached full production.

Although fighting an infection can take a toll on a healthy person's body, it is more likely to be life-threatening for very young children, whose immune systems have not yet perfected the swift manufacture of a strong response, or for the elderly, whose immune systems get weaker with advanced age. Infections are also much more dire for people of any age whose immune systems are already taxed by having to fight long-lasting battles against persistent, nasty infections such as HIV, or are being injured by treatments such as chemotherapy. In these individuals the immune system might make low-quality antibodies, or might not produce enough of the right kinds of killer T cells. A multiplying virus or bacteria can damage many cells in the lungs, kidneys, or brain, weakening them enough to cause serious complications that can kill. And often, additional damage is caused by an ineffective, poorly focused immune response that destroys innocent bystanders in its failed attempts to get the pathogen.

For the flu, the fatal complication is often pneumonia. Pneumonia occurs when an infection causes inflammation in the lungs; fluid seeps out of blood vessels and into air sacs, limiting their ability to take in air. The blood does not receive enough oxygen, and therefore can't deliver enough to vital organs. The kidneys begin to falter. The heart pumps like mad in a vain attempt to push more blood through the body to deliver more oxygen. Under these stresses the heart and kidneys can collapse, and the victim dies from a heart attack or multiple organ failure. Despite much research, there are still no cures for the flu.

Once your immune system battles a virus or bacteria, you generally will not have a problem with that pathogen again. Your body learns from the experience. It becomes immune to further infection. Once B and T cells have defeated an enemy, the survivors return to base camp. If, a month or a year or a decade later, they suddenly contact the same germ again, they don't need a week to figure out how to respond. They remember. The code for producing the best antibody has been imprinted on their genes. The battle-experienced helper and killer T cells spring into action. They go on the attack almost immediately, and tell new recruits how to do the same. The invading pathogen has little time to multiply, and is usually defeated quickly. You probably won't feel a thing.

This acquired immunity persists even though B and T cells eventually die. As they age they can divide, creating fresh mature cells with the same genetic memory. Still, the number of cells with memory may dwindle over years or decades. If bits of the original virus antigen settle in the body's nooks and crannies, they can be discovered later by new white blood cells emerging from the bone marrow. They will quietly send cytokines to base camp, and young B and T cells will extend the body's immunity.

The immune system's memory is a great asset. But it doesn't protect against every threat, especially when pathogens mutate regularly. The flu virus alters itself every year, which is why Nancy Cox and her CDC team must be so vigilant. If you are exposed to a major new mutation this year, your immunity from last year won't stop it.

If there were drugs to stop viruses, this issue might not be so problematic, but the pharmaceutical industry has had little success devising drugs that can kill viruses without also killing the healthy cells they invade. The best defense against viruses that the medical community has devised is the vaccine.

The British physician Edward Jenner concocted the first vaccine in 1796 to stop smallpox, then a highly contagious disease that wiped out entire towns. If a person got the disease and lived, however, he would never contract it again. Some Britons would purposely try to become infected with a "light" case of the deadly germ, in hopes they would survive and not have to fear the future. Many died. The survivors were left with pockmarked skin.

Jenner lived in the farming community of Berkeley, where people noticed that milkmaids rarely caught smallpox. They would even describe a woman who had a pure complexion by saying, "Her skin is as fair as a milkmaid's," meaning she didn't show the disease's signature scars. Jenner observed that milkmaids did routinely get cowpox sores on their hands, however. He theorized that the two diseases were similar, and that contracting cowpox somehow taught the body how to fight off smallpox.

Following his intuition, Jenner extracted a bit of pus from a cowpox sore on one milkmaid's hand and smeared it into a cut he made on the arm of a healthy eight-year-old boy, forcing the boy's body to fight cowpox. Six weeks later he brazenly injected smallpox into the boy's arm, in a sore left by the cowpox. The boy suffered no ill effects. After more experiments, Jenner formulated the first vaccine, a term he derived from *vacca,* Latin for cow. The practice of vaccination against smallpox rapidly became accepted across Britain, and laid the groundwork for future vaccines against many diseases.

By exercising the immune system, a vaccine increases the number of cells trained to kill a specific pathogen, and increases the speed with which they respond. It acts like a mock infection. A vaccine contains a minute amount of the pathogen or pathogen proteins in a small dose of fluid. A doctor or nurse injects the solution into your arm or backside. The pathogen may be dead, or attenuated—still alive, but altered in some way so that it no longer causes disease in healthy people. Newer vaccines sometimes use only selected pieces of the pathogen or even a selected gene. In each case, scientists have made the challenge weak enough so it does not trigger a full infection, yet potent enough to attract the attention of the immune guards. The immune system reacts as though it were under attack by the real thing. Macrophages or dendritic cells, the multifunctional phagocytic cells, "handcuff" the invaders and drag them to base camp. There, B and T cells analyze the germ's antigens.

If, later, you inhale the real virus or bacteria, the antibodies will block or slow down the invaders, the trained B and T cells will crank out more antibodies, and the killer T cells will deploy, stopping the infection before it causes harm.

Most vaccines are given in early childhood, and protect for decades or for life. But people must get vaccinated for the flu each year because it mutates. Once B and T cells devise a response that is precisely targeted to the invader, they will not effectively attack another pathogen that has even a slightly different target antigen. The flu virus regularly alters some of its antigens—the proteins in its coat that protrude from the surface—and that is enough to fool the system. The immune system has to reanalyze this new virus, which gives the flu time to rage.

When the changes are minor, the process is called antigenic drift, and it occurs regularly as flu migrates around the earth. If you are infected, but have battled a virus or vaccine the previous year, the news would not be entirely bleak. The old warriors would still put up somewhat of a fight while B and T cells prepared new defenses. You would still get sick, but would probably have milder symptoms than someone who had not confronted the previous strain.

That's what happens most years. What public-health officials dread most are the years when there are major changes—drastic antigenic shifts. Rather than alter one little protrusion on its surface, in such cases a flu strain swaps genetic material wholesale with another strain, changing each other's traits dramatically. This creates a whole new flu subtype that requires an almost entirely new assessment by the immune system, and an entirely new vaccine formula.

When such a case is detected—like the H5N1 chicken virus—Nancy Cox's team whips into high gear. They inject the mutated virus into some of the fifty ferrets that CDC keeps at its internal animal lab. Ferrets, it turns out, develop flu infections in a similar fashion to the way humans do, and their immune systems generate antibodies similar to our own. Cox's staff then takes serum—the clear fluid component of blood—from the ferrets and determines which antibodies are present. If the antibodies are different from any that have been created in the past by ferrets or humans, Cox can conclude that a genuinely new flu strain has emerged.

At the same time, Cox's crew sequences certain genes that create the protein coats of the suspicious strains, to see if they share a common lineage with known influenza viruses. "All of this," Cox says, "is to reduce the chance that we wrongly predict which flu strains will threaten the world."

Indeed, while the lab team is examining the odd viruses, Cox is talking until 6:00 or 7:00 P.M. every night with peers at the World Health Organization, the National Institutes of Health, the Food and Drug Administration, and the other world centers, all of which are trying to assess the myriad strains. "Working across the globe each day is very exciting, very rewarding," Cox says, with no sign of exhaustion. "The way we collaborate is a model for sleuthing other infectious diseases." Her struggle, however, is trying to stay ahead of the constantly changing virus. Although data comes all year, Cox admits that "we are still trying to read the tea leaves."

By December, Cox's group is constantly analyzing flu strains, but not for the sake of the current year's vaccine. Every December or early January the World Health Organization consults a panel of experts, Cox chief among them, about which emerging strains might imperil the most people worldwide. The group recommends three flu strains to be included in the following autumn's vaccine cocktail for the northern hemisphere. Three strains are enough to cover the most likely troublesome infections, and are also the most that manufacturers can mix into mass quantities of a single vaccine quickly and reliably. Each country must then decide whether to adopt the WHO recommendation, which is almost always done. WHO initiates a parallel exercise in September to kick off preparation of vaccines for the southern hemisphere's subsequent winter flu season.

Having to choose just three strains from the hundreds that circulate annually puts intense pressure on Cox. What if she fails to select a strain that subsequently roars around the globe? The year's vaccines would be useless against it. "I make sure I do the best job I can," she says. "Sometimes I wake up in the middle of the night and question myself. If we get it wrong, a lot of people will get sick. And a lot of other experts, agency leaders, even Congress, will be questioning us, too. That prospect keeps us on our toes—as if we needed more than a constantly changing virus to keep us on our toes."

In January, the FDA, which tracks what Cox and WHO are finding, informs pharmaceutical companies which strains should go into the new vaccine. The CDC provides seed virus for the three strains to the FDA, which tests them. By February the FDA sends the stock for mass production to interested manufacturers. The producers grow each strain separately, by injecting the stock into millions of fertilized chicken eggs. The eggs are incubated for several days to allow the virus to reproduce. They then draw off allantoic fluid (egg white) to harvest the virus. They purify each strain in huge vats, disable them with chemicals so they can't cause actual influenza when injected into people, and return them to the FDA for testing. They then blend the three strains with a carrier liquid into thousands of gallons of unified vaccine. They dispense the remedy a quarter-ounce at a time into millions of vials. If all goes well, by August they begin shipping vials to hospitals, clinics, doctors' offices, and other health organizations for arrival by October.

For the 2001–2002 winter flu season, the FDA recipe included antigens from the A/Moscow, A/New Caledonia, and B/Sichuan flu strains. Two of these strains had been included in the prior year's formula, but B/Sichuan was new. The three primary suppliers were Aventis Pasteur, Wyeth-Lederle, and Evans Vaccines, which produced 84 million doses for U.S. use.

Nonetheless, flu still swept across America in early 2002. As it does each year, it infected multitudes and killed 30,000 people. The pathogen still put approximately 110,000 people in the hospital, and caused 70 million work-loss days. And as always, the numbers worldwide were an order of magnitude higher. Why? Only a fraction of adults and children get vaccinated, even people over age fifty. And despite the vast experience of scientists and manufacturers, flu vaccines are only 70 to 90 percent effective in healthy adults—even less so in children, the elderly, and those already ill. So lots of people die.

This bare fact wears on Cox, but she plugs away. "We know a lot about the influenza virus. Yet there is a tremendous amount we don't know. That's frustrating. The virus can still surprise us. That's what causes pandemics. And preventing those is what keeps me going."

What scares her even more, though, is that the flu virus seems to be changing faster than before. "We know the next pandemic is going to come." She frowns. "In the summer of 2001 a strain we thought had died out in the U.S. and Europe after 1987 suddenly reemerged—first in Hawaii, then Canada, then the States. It had circulated in southern China for a decade, but then took off. We have no explanation for why." Cox gives a second example: "During the 2001–2002 winter, two influenza subtypes that had been circulating separately for years suddenly mixed together and exchanged genes—presumably when they met inside a single individual. We had had reassortment events like this in the past, but they didn't occur as quickly or as widely as this one. This worries us. It could be that the epidemiology of influenza is changing."

The 1997 Hong Kong outbreak and a smaller one that followed had already raised tremendously the CDC's worry about an imminent pandemic. In July 2000 it held a nationwide satellite video conference for doctors, public-health officials, federal, state, and local emergency preparedness planners, even funeral directors, on what they could do to lessen the impact of what was sure to be a coming outbreak. During the event, CDC experts said epidemiologists had agreed that the probability was high that another dangerous mutant strain of the influenza virus would evolve soon. Vaccine shortages and disruption of social and community services would be highly likely. Local emergency planners would have to implement large-scale triage centers and vast infection-control measures.

If we are to meet this threat, we must improve our capabilities for producing and distributing vaccine. Good science is not the only factor in producing effective vaccines. Once researchers figure out the year's best vaccine formula, the potion must be manufactured and distributed on a nationwide scale, and in recent years the system has been riddled with problems.

The flu season in the northern hemisphere usually begins in November. In the months leading up to it, manufacturers must produce, package, and dispense large quantities of vaccine quickly. A small delay can mean that vials won't arrive at health-care organizations until after the flu season begins, leaving people unprotected. Delay also means some organizations might cancel orders, resulting in financial loss for producers that claim they are not making much money on the work to begin with.

To ensure protection and profit, manufacturing has to proceed almost flawlessly. In recent years it hasn't. In 2000, the three makers, Aventis Pasteur, Wyeth-Ayerst, and Medeva, produced 75 million doses, sold in bulk at two to three dollars apiece. But delivery was very late—only 70 percent complete by the end of November. Low production yield of the A/Panama strain, and other production difficulties, were largely responsible. The situation forced health-care providers to give immunization priority to the elderly, children, and individuals with compromised immune systems.

Of course, blame was passed around. But responsibility had to be shared. In the January 2000 meeting that had been held among the CDC, the FDA, and company officials, the experts had agreed that A/Panama should be substituted for the A/Sydney strain that WHO had recommended, because the A/Sydney strain had grown too slowly in the past. The FDA basically sanctioned the substitution and provided the first seed virus. At the time of the meeting, at least, an FDA official said A/Panama showed reasonably high yield—fast growth rates. Nonetheless, A/Panama turned out to grow poorly for some manufacturers.

Even though certain manufacturers saw slow growth during the early runs, they proceeded to mass production. Perhaps test runs were shoddy, or, as one FDA report later suggested, the manufacturers didn't want to give up on A/Panama that late in the game because the process would have had to start all over again, with a different A/Sydney substitute. Production would have been even later, and health organizations might have canceled orders because they probably wouldn't have gotten the vaccine in time for the bulk of the flu season.

Unfortunately, lessons from this episode seemed to have been quickly lost. Delays occurred again the next year, leading up to the autumn 2001 flu season. Only two-thirds of the doses were available for delivery by the end of October. Once again government officials had to recommend that health-care providers phase in their immunization programs, reserving inoculations for high-risk people initially, and turning to lower-risk people in the following months as the remaining doses became available. Vaccine delivery in the fall of 2002 was better than in the prior two years, and there was actually a slight surplus.

Despite practical problems, vaccines remain one of the most powerful tools in medical science and public health. They have virtually eliminated some of the worst scourges ever to attack mankind. They have been successful for so long, however, that the American public is in danger of forgetting how tenuous the defense can be. We are dangerously close to a vaccination crisis in the country—not against flu, but against childhood diseases. Far too many parents are failing to vaccinate their children, and they are putting the entire population at risk. There is a quiet but sure resurgence of certain childhood diseases with names that sound as if they are relics from a distant past; supposedly eradicated diseases are returning. In 2000, more than 7,800 U.S. children were infected unnecessarily with whooping cough and 200 with mumps. Whooping cough has reemerged across Europe, notably in France, Germany, and the Netherlands, in what health officials there are calling a "hidden epidemic." In the Netherlands alone, 500 people are being hospitalized annually. Scientists suspect the reemergence may be due to new mutants of the bacteria.

We have not "beaten" childhood diseases. We must contain them actively year after year, generation after generation. Believing that measles or whooping cough has been eradicated is to flirt with disaster. The danger of this folly was proven in the 1970s: When British whooping-cough vaccination rates dropped below 60 percent, 100,000 new cases erupted; when vaccination rates dropped to 20 percent in Japan, 13,000 new cases arose. A measles epidemic shocked the United States from 1989 to 1991;[5] as many as 55,000 children contracted the disease and 132 died from it.

Not possible in your neighborhood? That's what residents of Douglas County, Nebraska, thought in the spring of 1999. Then an outbreak of rubella ripped through the community.[6] By August the disease had struck sixty-seven young adults and sixteen children. None had been vaccinated.

Parents who refuse to immunize their young sons and daughters against the gamut of childhood diseases are gambling with their children's lives. It is very difficult for a child's immune system to decode and fight these diseases. Throughout the 1990s, more and more parents chose to not have their children immunized. They claimed that the vaccines were dangerous, that they actually increased a child's chance of contracting a given disease,[7] or that they led to serious side effects such as diabetes or even autism. Some parents wrongly assumed the chance of their child getting one of the diseases was virtually zero. Others didn't want to introduce "those toxins" into their son's or daughter's body.

Certain mothers claimed they would rather pass on "natural immunity" through breastfeeding. Breastfeeding does pass immunity to babies, but it lasts for only a few months. Then the child's immune system must develop its own immunity anyway. There are only two ways to do this: get vaccinated, or face the actual disease. Breastfeeding offers no long-term protection.

Other parents do not immunize their children because they do not have insurance or the money to cover the shots, or they fail to take advantage of government aid. Still others don't bother to make sure their children receive booster shots when needed. Federal studies indicate that by 2000, fewer than 50 percent of U.S. children had received the complete regimen of recommended vaccines and boosters.

Furthermore, a study released in 2002 by the *Annals of Internal Medicine* found that individuals who had received their initial inoculation to certain diseases but skipped subsequent boosters had lost their immunity by adulthood. After doing bloodwork on 18,000 adults, researchers found that 40 percent were no longer protected against diphtheria and 28 percent were no longer protected against tetanus.

Many of the worries parents have about vaccines have no basis in science. Their views are rationalizations or, worse, dangerous myths. The views grow mostly out of paranoia or trendy skepticism about the medical community and the pharmaceutical industry. At the CDC's National Immunization Program, located on the same Atlanta campus where Nancy Cox works, it is Ray Strikas's job to educate the public about the truth. His steely eyes narrow and his speech quickens as he delivers straight talk about vaccinations, backed with volumes of no-nonsense facts.[8]

Strikas says there is always a chance that an individual will have an adverse reaction to an inoculation. But for most vaccines the risk is extremely low—on the order of three or four in a million. And most reactions are mild, such as a rash. The bottom line, he says, is that the risk of an adverse reaction is far less than the danger of contracting any of the childhood diseases, all of which can kill.

Childhood vaccines do not increase the chance that a child will contract the disease; they lessen it by training the immune system. In recent years the media hyped evidence supposedly showing that the standard measles-mumps-rubella (MMR) vaccine could cause autism. A coincidence between the time children are given the MMR vaccine and when the symptoms of autism first become apparent makes it seem to some parents that the two are linked. "This was based on a single report involving only thirteen children in the United Kingdom," Strikas explains, "and no one since has been able to verify or replicate this data." On the contrary, several large subsequent studies have shown no connection between the MMR vaccine and autism.

Strikas dispels other fears with facts, too. The hepatitis B vaccine can cause hair loss in a few individuals, he notes, "but it is only temporary. The hair quickly grows back." The disease, however, can last a lifetime. And he says there is no data to support the claim that a meningitis vaccine known as HIB can cause diabetes.

There is nothing inherently toxic about vaccines. This myth was started when people found out that thimerosal, a mercury compound, was used as a preservative in certain vaccines. Mercury in high concentration can cause neurological deficits. "But the level of mercury in an inoculation was about the same as that found in a can of tuna fish," Strikas says, "and thimerosal is no longer used." Even if thimerosal is present, a 2002 study confirmed that the risk is low. The study showed children rapidly excrete thimerosal after being vaccinated, so that it does not build up in the body.

Of course, vaccine production is not perfect. Certain children were made ill by an old-style diphtheria-pertussis-tetanus vaccine in the 1980s that was soon replaced, and a new rotovirus vaccine made by Wyeth-Ayerst was quickly pulled off the market in 2000 after problems were found. But these incidents are isolated and closely watched. People are at greater risk of side effects when they take even mild over-the-counter products for basic symptoms such as headache.

Perhaps the greatest misunderstanding is the belief that American children have little or no chance of contracting diseases such as measles or whooping cough. Strikas gets visibly aggravated at parents who confront him with that suggestion. "This is a free country and you can do what you want," he tells them curtly. "But you are playing roulette with your child's life." He stresses that parents have a social responsibility, too. "Your child has a better chance of not getting these awful diseases precisely because so many American children are vaccinated."

Many municipalities are getting tougher about parents' lax attitudes, and they are using school as the weapon. Increasingly, school districts will not allow children to enroll unless they have been immunized against certain diseases. The Massachusetts Department of Education, a leader, draws its list of required immunizations from recommendations by the state health department, the CDC, and the American Academy of Pediatrics. Most battles are fought each spring, when parents who are signing up their children for kindergarten for the next fall balk at the requirements. If parents don't have records—if their children are foreign-born or have moved from out of state—the children must be reimmunized. In Newton, school nurses refer such parents to Linda Walsh at the Newton health department.[9] "Often the parents fear side effects, but their fears are usually based on misinformation," Walsh says. "Other parents don't want the government telling them what to do." Although Walsh listens, she inevitably tells them flatly, "Childhood diseases really do kill children." Nonetheless, a few parents will not give in. And neither will Massachusetts. In these cases, some parents resort to homeschooling.

Some pediatricians are becoming activists on this issue, too, refusing to take on or continue to treat children whose parents refuse to vaccinate them. Others lecture parents sternly but still treat their kids because they are even more worried about the children becoming deathly ill. Bonnie Fass-Offit in Philadelphia is one such doctor.[10] "These parents are taking a ridiculous risk," she says. "It is hard for me to be their pediatrician. But our practice has had several children who have gotten very sick because they were not vaccinated. If no one helps them . . . then what?" If admonishment doesn't work, Fass-Offit asks parents to visualize one poignant scenario: "Imagine that your child has gotten one of these diseases. They are terribly sick. And they die. Imagine living with that guilt."

Fass-Offit's husband, Paul Offit, belongs to a nonprofit organization of doctors and parents that, among other things, has produced television-like documentaries to educate parents. The group, Parents of Kids with Infectious Diseases (PKids), was started in 1996 to inform the public about infectious diseases, the methods of prevention and transmission, the latest advances in medicine, and the elimination of social stigma borne by children who are infected with intractable diseases such as hepatitis C and HIV, for which there are no vaccines. The group has created a videotape that shows children who have terrible afflictions at home, and interviews their families to show the struggles they must endure every day.

At the CDC, Ray Strikas's insistence that children be vaccinated is based on statistics, not just social values. Basic vaccine coverage of American children has gotten better over the decades, but it is still in the low 90-percent range. Historical data from various Western countries show that childhood diseases can break out if more than a few percent of a population is not vaccinated. Measles, for example, can break out if more than 5 percent of children in a region are unprotected. In 2001, about 100 American children actually did get measles—most became infected while abroad and brought the measles back home—and another 100 got rubella. There have been no polio cases for years, but polio still exists in a few countries. All it takes is a single infected child traveling from another country to start an outbreak among U.S. children who have not been vaccinated.

Strikas says the danger is highest in certain geographic pockets. For example, in the area surrounding Medford in southwestern Oregon, inhabited largely by well-educated but defiant parents who seem to be outspoken against vaccines, about 30 percent of children are not vaccinated against measles. "They are ripe for an outbreak," Strikas says.

If statistics don't impress parents about the need for ongoing vigilance, maybe celebrities will. That's the reason the usually reclusive actress Mia Farrow has stepped back into the limelight. Farrow contracted polio when she was nine, and spent weeks isolated in a terrible polio ward where she heard adult victims scream and saw children die. Farrow has adopted numerous children over the years, including Thaddeus, fifteen, who was abandoned at a Calcutta train station, crippled with polio. She has spoken loudly in forums ranging from global summits at the United Nations to local town gatherings, to convince parents that no child should go unvaccinated—for his or her own health, and for the health of society.

Despite their phenomenal success, certain childhood vaccines actually now face commercial jeopardy.[11] The situation has become extremely tenuous. Even minor problems could cripple the entire U.S. system for making vaccines. The pool of licensed manufacturers has dropped from twenty-four in 1967 to four today. In 2001, shortages disrupted the normal administering of vaccines for eight of eleven preventable childhood diseases. The sporadic supply forced doctors across the country to postpone inoculations. In September 2002 some U.S. states began rationing the same vaccines because of ongoing shortages, and the U.S. General Accounting Office warned that future shortfalls were likely.

As with the flu vaccine, fewer and fewer manufacturers are interested in supplying the childhood mixtures because the financial return is low. Biotech companies now have the knowledge to develop clever vaccines, but few manufacturers can afford to supply them to the market. The government buys about half these vaccines at depressed prices, to distribute to clinics and other providers who give shots to the public at little or no cost. Meanwhile, the price of a full series of childhood immunizations has jumped from twenty-three dollars in the mid-1980s to more than $250 in the late 1990s. That has prompted more families to go to clinics, increasing the share of vaccines the government buys at a discount. Vaccine makers have little financial incentive to stick with the business, while at the same time they increasingly face lawsuits from people who claim to have gotten diseases despite being vaccinated or to have been harmed by the vaccine.

By the end of 2001, Aventis Pasteur was the only U.S. company left manufacturing tetanus booster vaccine, which created months-long shortages, forcing hospitals and doctors to abandon trying to give adults their usual booster shot every ten years.

Meanwhile, the company essentially stopped selling diphtheria-tetanus-acellular pertussis vaccine to government buyers, choosing to supply only private health-care providers at higher prices; this left British maker Glaxo SmithKline as the lone supplier of almost all public-sector vaccine. In 2002 Glaxo withdrew the only vaccine for Lyme disease because it decided there wasn't enough demand. Merck & Company was the only American maker of the measles-mumps-rubella vaccine, as well as the chicken-pox vaccine, which had only been introduced commercially a decade earlier. Although companies claim low profits as their disincentive, critics say manufacturers are not interested because they would rather sell a pill that people would take every day to combat a disease—an ongoing revenue stream—than supply a vaccine, a one-time sale.

The solution? Government must step in, says Philip Russell, professor emeritus at the Johns Hopkins University's School of Public Health, a former president of the Albert B. Sabin Vaccine Foundation, and now an adviser to the recently created Office of the Assistant Secretary for Public Health Emergency Preparedness. For years Russell has campaigned for a government-university-industry consortium to shore up research and supply. Others say the government must take complete control because industry is pulling out. Kenneth Shine, president of the National Institute of Medicine, has revived the call for a national vaccine authority owned by the government and operated by commercial contractors to crank out vaccines. Although industry has objected to this intervention before, Shine says the ongoing shortages, and the industry's poor readiness to make anthrax vaccine after the September 11, 2001, terrorist attacks, prove that industry has no business objecting. Independent analysts at the Defense Department, worried by incomplete supply to troops, recently came to the same conclusion: The government must make vaccines.

Massachusetts has taken the matter into its own hands. Way back in 1895, the state board of health established the Massachusetts Biological Laboratories to make diphtheria antitoxin. Over the next century the lab manufactured a number of vaccines for Massachusetts doctors that companies were not interested in supplying. The lab has produced a tetanus vaccine annually since 1941 for in-state use. But when the tetanus vaccine shortage developed nationally in 2001, the lab managed to produce an additional million doses for patients across the country.

Since then, lab director Donna Ambrosino has been getting a lot of calls from health officials, even investigators at the U.S. General Accounting Office, who are searching for an alternative to relying on the big pharmaceutical firms.[12] The lab is the only state agency in the country licensed to make vaccines.

But, regrettably, Ambrosino tells them, the unique effort may not be a model for other states. Her lab gets no state aid. It can barely break even only because of sales and royalties on drugs it developed over many decades. Ambrosino explains how incredibly expensive it is to devise and manufacture vaccines, and to operate a facility that continually meets FDA requirements. She adds that clinical trials and licensing take a long time and cost even more money. These are the very reasons pharmaceutical companies are getting out of the vaccine business. These are also the reasons why biotech startup companies are failing, Ambrosino says, frustrated. Like Shine, Ambrosino says the only way out is the "government-owned, contractor-operated" scheme, where governments take over such labs but let companies run them.

Despite the hurdles, Ambrosino is boldly expanding. In 2002 the University of Massachusetts, which runs the lab, announced it would build a new $77 million facility to widen the lab's unique role in producing vaccines, as well as "orphan drugs"—those that are effective against rare diseases but that generate too-little profit to attract drug companies. Ambrosino says the plan is justified because it will bolster public health: "These drugs would otherwise be substantially unavailable to the citizens of the Commonwealth."

More research could also spawn more-convenient vaccines. Shots have two disadvantages: Many people don't like them; and they don't always generate the best immune response. Oral or nasal preparations are more likely to mimic a natural respiratory infection and focus the immune response on the airways, where it is most needed. In 2001, biotech firm Aviron had finished phase-three trials of an attenuated flu virus vaccine that was a mist rather than a liquid. (Drugs must go through three trial phases that take several years before the FDA will approve them.) The mist could therefore be squirted into the nose by a nurse or, perhaps, by individuals in their own homes. A nasal spray would be less uncomfortable, and easier to administer than a shot. The product, FluMist, was up to 93 percent effective in preventing the flu, just as good as the best liquid vaccine. But some questions about the trials emerged in 2002, and in August additional information for review was given to the FDA by MedImmune, a company in Gaithersburg, Virginia, that had since acquired Aviron. In December 2002 the USDA approved FluMist, which was to be marketed by Wyeth.

Hoffmann–La Roche was pursuing a different, potentially more profitable flu therapy. In 2000 the firm began selling Tamiflu, a pill available by prescription. Tests showed that, taken daily, the pill shortened the duration of a person's flu symptoms by about a day, but, more important, reduced the symptoms' severity. As Hoffman-La Roche and the FDA monitored the drug's use in its first year, they found that if a healthy person took Tamiflu each day during a local outbreak, he would be less likely to get the flu in the first place. The FDA cautioned that the drug did not compare to vaccine in its potential preventive qualities, but was intriguing nonetheless. Of course, a cure would be the best news of all. And because millions of people catch influenza annually, a cure could pay off handsomely. Corporate researchers have been after this grail for years, yet no solution has emerged.

Clues may come from parallel efforts to cure the common cold. In December 2001, infectious-disease researchers from the University of Virginia announced they had developed the first medication proven to reduce the length of a cold, again by about a day, for infections caused by a rhinovirus, the most prevalent cold pathogen. Like the Tamiflu product, the drug interferes with the virus's antigens, which latch onto healthy cells and then bore in. The university's research was funded by ViroPharma Inc., which had begun petitioning the FDA for approval to market the drug in pill form. Another company, Agouron Pharmaceuticals, was in phase-two trials of a nasal spray providing similar results.

Because childhood diseases generally do not mutate, the standard vaccines prepared against them are sure defenses, if society can find a way to reverse the disappearing payoff that is driving companies out of the vaccine business. Government must step in and help industry fix the financial problem. To beat the flu, however, more research is needed. No matter how swiftly Nancy Cox's staff can chart the incoming flu strains, as long as our vaccines attack the flu's outer coat and that coat keeps changing, we will always be playing catch-up. A better potion—one that can disable the central core of the influenza virus and thus defeat them all—is needed. Scientists have not found it. And there is little research money available to help them try. Until we discover that solution, the pandemic will continue to lurk around the corner.

CHAPTER EIGHT

TB AGAIN: *The Fight-Then-Forget Cycle*

Of all the old scourges, none has proven more persistent than tuberculosis. And today TB is booming in part because it is developing resistance to multiple drugs, quickly learning to outsmart the concoctions we have to fight it.

TB is one of our oldest adversaries. It can be detected in mummies from 5,000 years ago. Hippocrates railed against the disease in 400 B.C. Physicians in the eleventh century called it the White Plague. Middle- and upper-class victims across nineteenth-century Europe trekked to Swiss sanatoriums in the vain hope that the alpine air would somehow clear their lungs of the disease. During the Industrial Revolution the disease, then referred to as "consumption," caused one in every five deaths in Britain, and in 1900 it was the second leading cause of death in the United States.

If you thought we had put TB behind us, think again. *Mycobacterium tuberculosis* (Mtb)—the bacteria's proper name—has evolved into one of the most virulent pathogens on earth.[1] It currently infects a stunning one-third of all human beings. That's 2 billion people! There are 8 to 12 million new cases annually, and the World Health Organization estimates that without aggressive intervention, nearly 1 billion additional people will be infected by 2020. What's more, TB has one of the highest infection rates of any disease in the world.

It is easy to become infected with tuberculosis. All you have to do is inhale a few invisible particles expelled by a contagious person nearby. People with active disease can spread the bacteria by coughing, sneezing, even singing or speaking. Five percent of people who are exposed develop the disease within two years, and another 5 percent do so at a later time. Symptoms are chills, fever, and night sweats that become progressively worse, a bad cough that persists for more than two weeks, chest pain, fatigue, weight loss, and sometimes coughing up blood or sputum (phlegm from deep inside the chest).

The other 90 percent of people exposed to TB remain carriers. Their immune system walls off the bacteria in molecular prisons in the lungs. This makes the bacteria inactive, but they remain alive. A person carrying this latent form of TB infection has no symptoms and can't spread TB to others, and in many people it will remain inactive for the rest of their lives. But in others the bacteria can erupt into active disease at any time—especially if a person's immune system lets down its guard as it is weakened by other chronic infections such as HIV, malnutrition, or simple aging.

Most often the bacteria grow in the lungs. When you breathe in, they attach to the alveoli (air sacs) there.[2] If you are lucky, your immune cells will soon find all of the particles. When a macrophage brushes up against a TB organism, it extends two arms and surrounds the bacterium. The arms fuse, forming a sac called a phagosome. This sac floats inside the macrophage, like a soap bubble inside a balloon. Another bubble inside the macrophage, called a lysosome, fuses with the phagosome into a larger bubble and dissolves the bacterium with enzymes.

Often, however, the macrophages do not succeed, because Mtb has developed a number of bold tricks that enable it to escape harm and multiply. The elaborate means by which Mtb can fend off immune system attacks are chilling examples of just how deviously clever pathogens are becoming. Rather than waiting to be captured—which sets off cytokine alarms—Mtb has evolved a way to actually sneak inside a macrophage. If it can slip in below the radar, the macrophage will be unaware the invader is hiding inside. Once there, the bacteria can freely multiply and infect other cells.

If the macrophage does detect the Mtb and stuff it in a sac, the bacteria can deploy several countermeasures. In one ploy, it manages to prevent the phagosome sac from fusing with the lysosome sac.[3] Normally the phagosome uses an enzyme called Hck that tells the lysosome to begin injecting the Mtb with lethal chemicals. Somehow the Mtb intercepts the Hck messages so they never deliver the death order. The macrophage's last resort is to generate new deadly peroxide and bleachlike chemicals. But Mtb can quickly produce at least three enzymes that can break down these caustic agents.

A macrophage is not too proud to ask for help when a tuberculosis cell stymies its attacks. It sends out cytokines calling helper T cells to come to its aid. The macrophage then rips off some of the Mtb's antigens. The macrophage's inspector, the MHC molecule, drags them to the cell's surface and pushes them through so that arriving helper T cells can analyze them and help the macrophage kill the enemy within. Here again, though, the Mtb can interfere. It produces a small, studded protein that prevents the MHC inspector from grasping the antigens. It can even stop the macrophage from producing new MHC molecules. The prisoner tries to take control.

If T cells do make contact with the Mtb antigens, they then help the macrophage by sending in the cytokine known as interferon. The interferon energizes the macrophage to destroy the Mtb in the sac. But the Mtb seems to know what's coming, and somehow blocks the incoming interferon signals. People who have genetic deficiencies that prevent them from making or responding to interferon or its partner, interleukin 12, suffer from severe forms of TB if they are infected.

At this point the macrophage has no other way to kill the enemy. Frustrated, the only thing it can do is try to prevent the clever bacteria from reproducing and breaking out to infect other cells. It can't achieve this alone. The macrophage again asks T cells for help. Sometimes the T cells will instruct the macrophage to undergo apoptosis—cellular suicide. The macrophage digests its own DNA and dies.

Apoptosis is risky business, however. If too many macrophages kill themselves, their host—you—will be left with a greatly weakened immune system. So, more often, the T cells accept that the best option at this point is a standoff. The T cells gather around the macrophage and transform it from a soft, round, balloonlike cell into a stiff, square chamber—a sealed prison. The T cells transform other infected macrophages nearby, push the prisons together, and wall them off with a heavy, fibrous coating of collagen. This creates a hard grain called a granuloma—a high-security penitentiary. In actuality, each granuloma is about the size and toughness of a grain of sand, which appears as a small dark spot on a chest X ray.

Once the many invading Mtb are encased in granulomas, they lie dormant, though still alive. The TB infection is now latent. It is still a grave danger, however, and everyone with latent TB must be treated. That's because the imprisoned Mtb is down but not necessarily out. If the immune system starts to wane, the bacteria can break out. They wait for an opportunity to slip past T-cell guards that patrol the penitentiary. Perhaps the cells are called to fight another persistent infection or have been killed if the human host has contracted HIV. Or the guards may simply weaken due to poor nutrition or old age. Mtb also has a trick that can sometimes destroy the guards: It induces infected macrophages to send a death signal to T cells that get too close.

When the immune system is sagging, Mtb can break out, boring through the granulomas so it can go on its long-awaited rampage. It multiplies fiercely in surrounding alveoli, clear-cutting swaths of lung tissue, creating blank cavities in some places and cheesy mush in others, which appear on X rays as white bare spots and blackened sludge, respectively. Both types of destruction limit the lung's ability to absorb oxygen. Meanwhile, the destruction fills the lungs with waste fluid, drowning other alveoli. Eventually, if left untreated, the victim can suffocate and die. In addition, the infection can cause the body to waste away or consume itself—that's where the old name "consumption" comes from—raising a victim's vulnerability to other infections as well.

Chest X rays and sputum tests readily show evidence of active TB. A skin test is the simplest way to find out if you have a latent TB infection, however. A nurse injects a small amount of fluid that contains a TB antigen just under your skin, usually on the lower part of your arm. If a significant hard bump develops after two or three days, you likely have latent TB. It is a sign that your lymphocytes have met the antigen before.

Physicians have only five primary drugs that can effectively vanquish TB, but the bacteria can show resistance to all five. There is a secondary group of drugs doctors can use if the first set don't work, but they are more toxic and less effective, and TB is beginning to develop resistance against them, too.

If you have latent TB, the usual treatment is to take isoniazid for six to nine months or more. It penetrates the granulomas and kills the imprisoned bacteria. People with active TB must take a cocktail of some or all of the five prevailing drugs for at least nine months and as long as two years if they have a resistant strain, to overcome the fierce TB as well as to overwhelm its potential drug resistances. The five first-line compounds are isoniazid, rifampin, pyrazinamide, streptomycin, and ethambutol. Streptomycin was the first, discovered in 1944. But by 1948 resistant strains had emerged. Isoniazid and pyrazinamide were introduced in the 1950s, and the others later. Common side effects of the various drugs are loss of appetite, nausea, fevers, and abdominal pain. Rifampin can turn urine, saliva, or tears orange, and can make your skin very sensitive to the sun.

Complete killing of TB in any patient requires lengthy treatment. If you don't take your medication to completion, some bacteria will survive, and potentially break out from their prisons, reproducing into newer forms with tricks to resist the drugs. Your only recourse, then, is second-line drugs that must be taken for a longer time, have more serious side effects, and simply don't kill Mtb as effectively. In the interim you can become infectious again, and can give now-drug-resistant TB to your family, friends, or anyone you spend time with.

Although we may think that drug-resistant TB must surely be a threat only in the so-called third world, it is also a serious problem in many developed nations. In certain parts of the Russian Federation, as many as 22 percent of new TB cases are multi-drug-resistant, and nearly 80 percent of patients in some Russian prisons, where the disease runs rampant, are resistant to at least one drug. In 2000 alone, multi-drug-resistant infections doubled in New Zealand, increased by 50 percent in Germany and Denmark, and leaped by 80 percent in London. Latent TB is carried by nearly half the population in China, and infects 38 percent of people in Japan, where it has even infiltrated the household of the imperial family's heirs.[4] Worldwide, 60 million cases are drug-resistant and the number is rising. If untreated, 70 to 90 percent of people with active multi-drug-resistant TB die, half of them within four to sixteen weeks, as the organism steadily destroys the lungs.

Although TB is less common in the United States, we are increasingly vulnerable, owing to high levels of international travel and relocation. Nearly 40 percent of U.S. TB patients are foreign-born. America's cultural mix is changing, and so is its mix of infectious diseases. The problem had become serious enough by 2000 that officials at the Institute of Medicine called for increased TB screening of new immigrants coming from Mexico or any country where more than 35 percent of the population was infected.[5] They proposed that people with latent infections could still enter this country, but would have to be treated before being given permanent-resident status. Britain and Australia are considering similar measures.

Multi-drug-resistant disease is breaking out in local communities all around us. In Virginia, for example, the number of cases more than doubled from 1999 to 2001. And although it's true that people who live in crowded, unsanitary conditions are more likely to become sick and infect others, TB does not discriminate.

In September 1993 one girl at La Quinta High School in Westminster, California, spread drug-resistant TB to 292 students in two weeks.[6] In 1999 one contagious member of a church in West Midlands, England, infected sixteen others after attending only two Sunday services.[7] In March 2000 a teenage boy working at a McDonald's in Fribourg, Switzerland, spread TB to coworkers. In February 2001 in Devon, England, nine regulars at the local Heavitree Arms pub became infected with TB in only a few days.

And recall that only one week after the two English epidemiologists, Philip Monk and Gerry Bryant, announced they had solved the mad cow puzzle, they had to battle a huge outbreak of tuberculosis at a school right next door to their Health Authority office. Of the school's 1,200 pupils, 251 were infected with latent TB, and seventy-one had active disease, an unusually high percentage. It was the most concentrated outbreak in modern England's history.

TB has reemerged in developed countries for one reason: Public-health agencies let down their guard. Rather than stamp out TB, they allowed low levels to smolder, which gave the organism time to develop stronger drug resistance and become a much tougher foe to fight. The disease first began to yield after World War II as effective drugs became available. For three decades public-health personnel aggressively located and treated TB victims. By the late 1970s, TB was virtually eliminated in developed countries.

Believing they had won not just battles but the whole war, health institutions in the United States and other developed countries had blithely scaled back their work by the mid-1980s. City and state health commissioners dismantled programs to track TB cases and to ensure that people with active disease finished their long course of prescriptions. By failing to take their medication properly, or by stopping early, patients helped the bacteria develop drug resistance. Then they spread the intractable strains to others. To make matters worse, millions of people became infected with HIV, which weakened their immune systems and made them many times more likely to succumb to TB.

As a result of assuming victory prematurely—on personal and public-health levels—the number of annual TB cases in the United States surged to 27,000 by 1992. In 1984 only 10 percent of disease cases worldwide resisted one of the five prevailing drugs, but by 1995 a daunting 52 percent were resistant to at least one drug, and 32 percent were resistant to two or more in some hot spots. The WHO became so concerned about the modern TB epidemic that it took the unprecedented step of declaring it a global emergency.

The U.S. public-health system has a checkered history of preventing TB. In the centuries before drugs became available, the plan was largely to stop foreigners who might be carrying the disease from entering the country, and otherwise to isolate victims in quarantine until their immune systems cured them or they died. Once drugs were available, public-health departments tried to find and treat as many people as possible. But as the numbers of cases declined, so did their vigilance.

As Laurie Garrett notes in her book *Betrayal of Trust,* this fight-then-forget cycle began in the United States soon after the dawn of public health itself, and it undermines our ongoing prevention of many infectious diseases.[8] In 1805, New York City created the nation's first board of health, in large measure to clean up filthy water that bred yellow fever, which was decimating the city's population. The business community widely supported the board, which had a first budget of $8,500. But the agency became a victim of its own success as it brought yellow fever under control. By 1819 its budget had been cut to $500. Sure enough, a yellow fever epidemic tore through the city in 1822. This pattern, Garrett says, continues to this day and contributes to what some call an American public-health system that is in disarray.

New York City's health department, one of the strongest in the nation, failed to learn a lesson from its own history. As TB faded in the 1970s, it cut funding for TB clinics and moved the money over to fight the rising tide of HIV. By 1985 it had reduced its TB clinics to the point where it was difficult for patients to get help. It scaled back on finding victims and tracking them to make sure they completed treatment. HIV was zooming. The health department needed money from somewhere, and it was costly to run TB clinics. The shift in attention was partly political, too. Most of the victims at that point were foreign-born adults, drug addicts, or poor, unemployed individuals.

In less than five years TB returned with a vengeance. The rate of disease cases tripled. But the city's health infrastructure to fight it was gone. TB clinics couldn't be instantly created out of nothing. The many HIV patients made the situation even worse; they often succumbed to TB and were concentrated in certain hospital wards, which caused TB and the drugs against it to be passed around in confined areas. That allowed multi-drug-resistant TB to spread and take hold. Follow-up treatment was nonexistent. A famous study in 1991 of a group of patients with active TB at Harlem Hospital who were discharged showed the health department could account for only 11 percent of them getting the follow-up treatment necessary to cure them.[9] The researchers had no idea what had happened to the other 89 percent. By 1992, TB raged, and the city once again had to spend enormous sums in an emergency effort to beat it down. TB is like a forest fire; it may start small, but suddenly the wind kicks up and the fire blows through the whole forest.

Unfortunately, New York City's failure was not unique. John Grange, a visiting professor at University College in London, says the same phenomenon is happening in London. From 1988 to 1998, TB cases rose 21 percent in England and Wales, with London reporting a 71-percent increase.

When New York City's health officials were criticized in the mid-1990s for dismantling their TB-prevention system, they cried that they did not have the resources to keep it going and also fight the rise of HIV. But in the end it cost the city more than $1 billion to bring the new epidemic under control; studies show it costs ten times as much to cure a person with multi-drug-resistant TB as to cure a person with regular TB. Other municipalities made much smarter decisions. Boston is a stellar case in point. If the United States, or any country, hopes to beat down multi-drug-resistant TB, it must learn the lessons Boston's story provides.

Unlike New York City, Boston had only a modest rise in TB in the early 1990s, and had no increased proportion of multi-drug-resistant disease. The secret, says Al DeMaria, assistant commissioner of the Massachusetts Department of Health and head of its Bureau of Communicable Disease Control, was that the city preserved the basic components of free clinics for treatment, public-health nursing, linguistically and culturally appropriate outreach, superb laboratory support, and inpatient treatment.[10] "We had scaled back some from earlier years," DeMaria says, "but we kept every component. We kept accessible clinics. We provided free medications and handed them to every patient so we knew they took them." Boston's fine system is still going strong. More than

98 percent of its patients complete treatment, much higher than the average for other cities or states. "It's a difficult system to maintain," DeMaria says. "The drugs have gotten more expensive. Testing is more expensive. But it actually works extremely well."

Boston's program shows that the key to beating TB is heavy reliance on a practice known as directly observed therapy, or DOT—giving each person each dose of drugs face-to-face. The reason personal delivery and supervision of treatment is so important is that many patients fail to take their drugs for the full six months or more required to truly eliminate the disease from their bodies. In the beginning stages, a patient feels so sick that he'll do pretty much whatever his doctor tells him to do. But after a few weeks of taking drugs, patients feel so much better that many either stop taking their medication because they don't like consuming many pills a day, or because they have uncomfortable side effects. Some patients simply forget, and skipping a few doses can allow the disease to gain strength. Studies show that in cities that institute the DOT system, the rates of TB, and particularly drug-resistant TB, drop substantially.

For patients in Boston, the ordeal of treatment begins at the third-floor TB clinic in Boston Medical Center's outpatient building, a massive facility that covers an entire square block in the south-central zone of the city. There a melting pot of patients stare blankly at big posters that say "Stop TB" in English, Spanish, Chinese, and Vietnamese, and wait for their names to be called. Some of them casually step into a room just to take their pills and emerge five minutes later, a routine stop on their way to work. Others look worried. They are the people about to begin their six- or nine-month odyssey of treatment, and they fear for their lives. Inevitably, they are the ones called into a room labeled "TB Triage Office," where they will be told they need a chest X ray. If doctors think they are contagious, they will be masked, placed in an isolation room, and may be admitted into the hospital until results are back. If disease is suspected they will be assigned to one of several possible treatment regimens.

Some patients are asked to come to the clinic at the same time every day so that a nurse can make sure they take their medication. Others, the nurses judge, can be trusted to follow strict instructions to take a certain number of variously colored pills at precise hours every day, and are sent off with a month's supply. If, however, the TB staff are worried that an individual's disease is advanced or is drug-resistant, or if they fear the patient might not reliably take their medication or show up for appointments, they assign a DOT nurse to bring the medication to that patient every day. This arrangement can be much more difficult than it might sound because, suddenly, a nurse is physically inserting himself or herself into a person's private life.

The degree of commitment to effective treatment among the staff of the Boston clinic is inspiring, as is exemplified by one case handled by nurse Michael Malone.[11] The case also highlights just how fraught with complications the battle against TB is. Malone, a thin forty-three-year-old, is one of the clinic nurses who evaluates new patients and administers drugs. But the heart of his job is to visit TB patients at home or work, give them their daily or weekly pills, and stand there while the pills are taken. It sounds like a simple procedure, but the process is replete with medical and social difficulties, not to mention its cost; if public-health systems are to conquer TB, they have to ensure that there are enough nurses to supervise patient treatment.

At eleven-thirty one November morning it was time for Malone to head out to meet his twelve-o'clock patient, Julia Corazon (not her real name). He grabbed his blue jacket, took the elevator to the parking garage beneath the hospital, got into his scratched-up Mazda 626, and drove to the rugged southern edge of Boston's Chinatown, not far from the office of Stuart Levy, the drug-resistance crusader. He parked at a meter alongside a city highway and locked The Club onto his steering wheel to prevent thieves from stealing his vehicle. He walked the four windy blocks to Julia's workplace, a set of offices on the fifth floor of a small, reconditioned office building on a narrow street veiled by the constant shadows of high-rises that surrounded it. The little company Julia worked for trained women who were on low income or public assistance how to become certified to take children into their homes for day care.

Julia, thirty-four, grew up in Ecuador. When she was sixteen she got very sick. Doctors incorrectly told her she had pneumonia, and made her take little red pills. She emigrated to Boston in 1986, when she was nineteen. She got a job and built a modest life. Seeming to be asthmatic, she had chest X rays in 1995, 1997, and 1999. They all showed the same scar on one lung, but it was stable. By the spring of 2000 she was feeling particularly ill. Her doctor sent her to a pulmonary specialist. He took an X ray and saw that the scar had changed. He took a sputum sample and sent it to the lab. Julia tested negative for tuberculosis.

Nonetheless, the Spanish-speaking pulmonologist suspected TB and told Julia's primary doctor to put her on four different drugs from June to August. Her sputum tests remained negative, but Julia's doctor told her to keep taking the medication. By December 2000 she felt measurably worse. On January 20, 2001, the pulmonologist requested a special TB lab test, and it showed what he had feared: Julia had TB. Further testing over the next four weeks indicated it would resist multiple drugs. The doctor told Michael Malone.

Worried that Julia had not taken her medication as directed, which would have allowed the TB to develop drug resistance, Malone interrogated all the prior doctors. They assured him she took her pills. Malone challenged them: "If she was taking four drugs for six months and her TB got worse, then the medication wasn't working, was it?" That meant Julia could have had multi-drug-resistant TB from the start. She could have been spreading the fatal infection like wildfire.

One doctor told Malone that Julia was married, had two children, and lived in a crowded section of Jamaica Plain, a struggling haven for South American and Caribbean immigrants on Boston's south side. Her mother lived with them, too. Now Malone had to perform a dicey duty—he had to meet Julia face-to-face, explain her dire situation, and assess why her medication was not helping. He called her at home and a man answered. Malone knew not to say, "Hi, I'm from the health department." Health workers have learned from bad experiences that spouses or boyfriends may not know that their partners have TB, and can become violent if they suddenly find out. But Malone had negotiated tricky situations for seven years, and he talked his way to Julia. She gave him her address.

The next day Malone arrived at Julia's apartment at a brown multifamily house squeezed between similar structures in an alley behind stores lining a main avenue in Jamaica Plain. He didn't know if she would meet him at the door, or if her husband would loom over him as if all this were a scam to rob them at gunpoint. But Julia answered his knock and politely showed him in. Her light brown face was strikingly beautiful, with a broad, quiet smile beneath a wave of auburn hair. She stood straight, facing her thin, five-foot-two body squarely at Malone, when she told him in her Ecuadorian accent that she had indeed been taking her medication. She took the pills at various times during the day, because swallowing them all at once nauseated her. Malone winced. He commended her for sticking with a tough regimen for so long. But he told her in his caring, understated voice that she had to take all the pills at once, so the drugs would spike high enough in her body each day to kill the TB. By keeping a constant low dose she was not only not challenging the bacteria, she was unintentionally helping it learn how to resist the medication. Julia's head dropped. She almost cried.

The doctors changed Julia's regimen. They put her on intensive therapy for fourteen days, during which time she had to stay at home. More lab tests ultimately showed Julia did not actually have drug-resistant TB, but a particularly intractable variation. "We dodged a bullet," Malone told her. Then she was put on DOT. From January 24 till April 3 he met her at her office every day and gave her the four drugs. Seeing good test results, the doctors then reduced her to two drugs twice a week.

As Malone crossed Chinatown on this Monday, the date struck him; it was November 5, nearly the end of 2001, and Julia would soon be off the drugs. She would have her final X ray, and, he hoped, be cured. He often subconsciously reviewed Julia's story as he walked past the Wang Center and crossed Tremont Street to her office. It spurred him on. His long hours, the late-night calls when someone arrived at an emergency room raging with disease, the uncomfortable home visits, the sometimes threatening patients, and the tiresome deliveries—Julia's success story made it all worthwhile. He greeted the receptionist at the child-care training office exactly at noon. The receptionist said nothing, familiar with the routine, though Malone had never actually explained why he was there.

Knowing Malone would be punctual, Julia immediately appeared from around the corner, in a sharp white blouse and pressed black slacks. She said little as she led Malone to an empty office along one corridor. Inside, Julia had set two paper cups full of water on the vacant desk. Malone chatted as he pulled a small manila envelope from his pocket, tore it open, and let three white and two red pills tumble into her palm. Julia took them slowly, finishing one cup of water, then the next, as they spoke purposefully. Julia's mother now had latent TB. They agreed Malone would arrive at her apartment at seven o'clock the following evening to see her. Malone asked Julia if he could bring his young son along, so he wouldn't have to find a baby-sitter, and she agreed. By now the two households were somewhat like family. Five minutes after Malone had arrived, he was gone.

Malone would make 1,000 such visits to twenty patients in the coming year. So would four other Boston health department nurses, to cover the whole city.

Back at his car, out of the November wind, Malone explained that Julia had been uncommonly accepting of his role. Others disdain him as the heavy hand of the state, treating them like children who cannot be trusted to take their medicine. Some are illegal aliens afraid he is an undercover agent for the Immigration and Naturalization Service. He tells them he won't report them, and yet they will meet him only in an upstairs room of some abandoned building so he can't pin an address on them. Julia, he says, is responsible. "She would take the drugs on her own." But because Julia had taken the low dose for so long, Malone had been worried about hidden drug resistance. He worried she might skip a dose, too, if she felt particularly nauseated on a given day. Furthermore, he explains, "when you drop to taking meds only twice a week, if you forget one dose you've gone seven days with no medication. That can create too much of a dip, allowing TB to rise again." Julia might say all this is a burden, he admitted, "but she is committed to seeing it through. She has the desire to get better."

Because of its commitment, Boston maintains a 98-percent treatment completion rate. The rest of the United States must heed Boston's call. The boom-and-bust cycle does not work. Although the number of people who contracted TB in the United States in 2000 was lower than the peak years of the early 1990s, there were still 16,377 new cases. No city or state can let up, because the new infections keep coming, due to unceasing international travel and immigration. Until TB is stamped out worldwide, it will always be ready to break out in America, Europe—anywhere.

Challenging as is the effort to fight the disease in the United States, the obstacles faced in developing countries are even more daunting. The good news is that since the early 1990s, a version of DOT known as the Directly Observed Therapy—Short course (DOTS) has saved millions of infected patients. That year WHO developed a strategy for getting TB under control, which included standardized DOTS surveillance and testing for drug resistance. Although its goals for 2000 were not met and had to be pushed ahead to 2005, it made great progress. Lack of money is the greatest hurdle; WHO estimates that it will cost $1 billion each year to treat the developing world, where most of TB is located and resources are scarcest, and another $200 million a year to treat low- and middle-income nations.[12] The Global Fund to Fight AIDS, Tuberculosis, and Malaria has raised $2 billion from governments, corporations, and individuals such as George Soros and Bill and Melinda Gates to make up the difference between what organizations like the WHO can afford and the real costs they confront. Yet even with slipped deadlines and limited funds, DOTS programs are producing cure rates of up to 95 percent in even the poorest countries. In parts of China, cure rates, among new cases anyway, are as high as 96 percent. In Peru the success rate is 91 percent. The effort is also preventing the development of more multi-drug-resistant TB by ensuring that the full course of treatment is followed. The World Bank has ranked the DOTS strategy as one of the most cost-effective of all health interventions.

Nonetheless, many people still go untreated. The challenges presented by the international scope of the TB problem are formidable, and to some degree represent a vicious circle; TB rates may decline in a country, only to rise again when people who are infected travel there with the disease. The degree of vigilance required to keep people who are carrying infectious TB from spreading the disease is demonstrated by another intriguing case handled by Malone.

Suzie Nackiwanda (not her real name) worked for a computer consulting company in Uganda. In March 1998 she flew to Boston to attend a conference she thought would help her in her job. But she never made it to the event. She felt weak and was coughing relentlessly. The tall African native, thirty-nine, turned up at Boston Medical Center, where emergency-room doctors took a sputum sample and X rays. Three days later they contacted Malone. He called Suzie in and told her she had pleurisy—a condition in which the membranes enveloping the lungs become swollen and congested. He also suspected TB and HIV. Suzie spoke clear English—the official language of Uganda, used to bridge the many tribal tongues like Suzie's Iutoro. Malone tried to get a clear story from her about the supposed conference, because her medical chart said she had stayed in a shelter in Boston. She gave him no American contact, just a phone number for her brother in Uganda. Malone admitted Suzie to the hospital for ten days and put her on four TB drugs while they waited for lab test results. She eventually produced a passport and a six-month visa, and told Malone that her sister had died of TB some time ago. Malone suspected Suzie had known back home that her life was in jeopardy, and that she would not get good treatment in Uganda, so she had come to the United States in a vain attempt to save herself. She'd left her sixteen-year-old son with her brother, and her twelve- and two-year-old sons with another sister. Her husband had died in a car accident in 1995.

Regardless of her story, Malone would not let Suzie out of the hospital. The tests showed she had raging TB, and could readily spread the disease to others. She probably already had done so during the two weeks she had wandered around the city lost in her predicament. It was the health department's mandate to protect the people of Boston, so doctors would treat her.

Malone told Suzie that because she was officially homeless, he would have to admit her to the Lemuel Shattuck Hospital, a state-run facility that has a TB unit on a locked floor. The big brick structure in southern Boston sits alone, cordoned off by twelve acres of woods. She said, "Go ahead. Do anything. I am sick and don't want to be." He put her in the hands of the Shattuck's doctors in the specialized TB ward, one of the basic elements of TB control that Massachusetts had preserved, which most other states had closed down in the 1970s and 1980s.

Suzie suffered a daily cycle of fever and chills for seven weeks—far too long, in Malone's opinion. Her TB was resisting the drugs, and her body was racked by the ordeal and the mounting infection. The doctors finally put Suzie on a cocktail of second-line drugs. Slowly she regained her strength, and the doctors started her on HIV medication because she had that disease, too. After six months of struggle, Suzie had recovered enough that doctors could no longer justify keeping her at the Shattuck. He told her she had to take drugs for eighteen months to be completely cured. Stopping now, half-treated, was dangerous; the TB could easily return. If she refused, she would have to return home, and risk a relapse as well as infecting her children. Suzie panicked. Her visa had expired and she was now an illegal alien. She wanted the treatment, but had no money and nowhere to stay. Malone said she could call Immigration to get approval to stay for treatment as long as she continued taking her meds. But Immigration could say no, too. She decided not to call.

Suzie could have walked out Malone's door right then. But if she did, Malone could have the health department declare her a menace and have her detained, although he knew, as he told her this, that she could quickly disappear into Boston's hidden alien community. Suzie wanted to get well, but did not want to stay at an awful shelter and did not want charity. She was eager to work. Malone made a call to a couple who ran a Christian community house for refugees in nearby Dorchester. They quietly took Suzie in. A social worker got her a job cleaning a house, and Suzie parlayed that up to her own list of ten clients. She told people she was in the country for medical treatment, so her questionable alien status never came up. But she worried constantly that there would be a knock on her door that would deport her home.

Suzie stayed six months at the community house, but the couple became too strident, requiring her to pray multiple times a day even though she had told them she was not a Christian. They berated her for refusing to be "saved," and told her she would have to leave. Again panicked, she turned to Malone. He got her an apartment in a building subsidized for AIDS patients, since she did have AIDS. She didn't have enough money to get in, so he put up her first month's rent. The building's managers said she could stay as long as she went to the meetings for residents with addictions and social problems.

In November 2000, Suzie finished her treatment and wanted to go home, concerned that her boys were growing up without her. Suzie would still need drugs to fight the HIV, and she worried she would not get them in Uganda. If she didn't, the HIV could weaken her immune system, and her TB could maybe reactivate. She could once again infect more people in Uganda and in the United States if they, too, traveled there, jeopardizing many lives.

Suzie landed in Kampala and stayed; her brother was nearby, and jobs were more plentiful. She shortly arranged for her children to move to the city. Her baby was now a walking, talking child who did not recognize or treat her as his mother, which hurt her. But she would start the relationship anew. Today Malone agrees he went far beyond his professional responsibility to care for Suzie. But, he says, "I wouldn't be a nurse if I couldn't go to an extreme for an extreme case. Suzie never asked me for any extra help. That is what encouraged me to do more."

A year after her return, Suzie, speaking from Kampala, was still glowingly grateful to Malone. "He saved my life. And he changed my perspective on Americans. Here, people think everyone in the United States is cold, that they won't even say hello to you on the street. But Michael—for him to do what he did for me . . ." She chokes up. She also says the persistence of Malone, the social workers, and others is what really cured her and prevented others in Boston and Uganda from contracting deadly TB. "With the fevers at the Shattuck, the change of drugs, the shelters—without the medical workers I would have gotten fed up and would have gone home." It was unclear, however, whether Suzie was getting the HIV drugs she needed, jeopardizing her immune system's ability to keep TB in check. Thus far, the TB had not returned.

The lengths Michael Malone went to treat Suzie and Julia are inspiring, but they are also eye-opening. If it takes that much to cure two individuals, imagine the effort to treat 2 billion people worldwide. Fortunately, a strict regimen of medication can cure many of them, and there are some positive signs that with the appropriate concentration and devotion of resources, TB can be brought under control.

Many public-health officials once thought multi-drug-resistant TB was virtually untreatable in developing countries because those countries had such inadequate health infrastructures and so little money. This attitude has been changed by Partners in Health, a nonprofit organization based in Cambridge, Massachusetts, and headed by Harvard physicians Paul Farmer and Jim Yong Kim. They were able to achieve cure rates in poor communities of Peru that were as good as or better than those reached in U.S. municipalities, even though they were treating multi-drug-resistant TB. Together with Harvard Medical School and Doctors Without Borders, they negotiated with pharmaceutical companies to reduce the cost of a full treatment of alternative drugs needed to beat multi-drug-resistant TB from $10,000 to $20,000 per person to less than $1,000.[13] Farmer received a MacArthur Foundation "genius" award, and Doctors Without Borders got a Nobel Prize for their long-standing work providing health care for the world's neediest.

Of course, if the world is ever to really wrestle TB to the ground, it will need new ways to counteract drug resistance. Like staph and strep, TB's entrenched resistance to multiple drugs does not leave physicians with many chemical alternatives. For years microbiologists have been hampered in developing newer drugs because Mtb behaves differently when grown in lab cultures than when it is living inside the human body. When they found an occasional new chemical that killed Mtb grown in the lab it was not effective in lab animals. Experts are at a loss to explain this inconsistency. At first they thought the compounds were unable to penetrate the granulomas, but that turned out not to be the case. Now they think that when TB is in a latent state in the body, it has a different cell chemistry than it does when it is active, and the bacteria in a lab culture are actively growing.[14]

The standard alternative research approach would be to test drugs in lab animals that have latent TB. Mice are commonly used in research because their immune system has been studied extensively, and for the most part behaves similarly to the human system. But TB advances somewhat differently in mice than in men.

Because the classical lab culture and lab animal approaches to developing drugs that can overcome resistance have not panned out, TB researchers, like many other disease specialists, are turning to genetics. Microbiologists finished sequencing the Mtb genome, which has about 4,000 genes, in 1998. They are now searching for clues to how Mtb works, and what its weak spots might be. Early studies have looked at which genes seem to be active when Mtb is trapped in granulomas, yet may be inactive when TB is growing in a lab culture. Scientists have already disabled several of the novel genes, creating a modified bacteria that no longer causes disease in mice. The next step is to develop drugs that target these genes.

Investigators at the University of Lucknow in India and at Hunan Medical University in China are trying to improve treatment in a different way: changing the delivery system for existing drugs, most notably isoniazid, by developing an aerosol that can be inhaled.[15] If a patient can inhale a drug directly into the lungs, they theorize, it will have the maximum effect on infected macrophages, which could shorten treatment and thus improve the chances that patients will complete their regimens.

In another effort, the National Institute of Allergy and Infectious Diseases (NIAID) is funding research into several possible compounds that might provide novel ways around TB's resistance. One possibility is to create a different class of TB drugs by modifying thiolactomycin, a compound that blocks the synthesis of certain components the bacteria needs to form its cell wall. If it can't build the wall, it can't grow and reproduce. Because the process is common to many microbes, it may be a promising drug for other diseases such as drug-resistant staph and enterococcus infections, as well as malaria.

Other scientists are after better TB lab tests. Looking for TB in sputum under a microscope is quick but not very accurate; it misses a lot of cases of active disease. Trying to grow the bacteria in culture is more accurate but very slow. Tuberculosis takes three to eight weeks to grow in a standard lab culture, unlike the typical twenty-four hours for most bacteria. It is impossible to put every person who walks into every emergency room or doctor's office and appears to have TB into an isolation room for weeks until test results are in. Yet if a patient is contagious, he can infect numerous other people while waiting for the results. Several companies have recently devised "rapid tests" that can indicate TB in four to six hours, but the tests are expensive. That kind of quick turnaround, however, would help doctors persuade a suspect patient to stay in hospital quarantine during testing, or would at least get the patient back into treatment sooner, reducing contagion. Mike Barer, at Leicester University in England, has developed an experimental blood test based on markers associated with the Mtb genome that takes only two to three hours. Barer tried it on students exposed to a recent outbreak there and is evaluating its performance.

The world does have one vaccine against TB. It is an attenuated form of TB, bacillus Calmette-Guerin, referred to as BCG. It is reasonably protective for children, but provides only minor protection for adults. It does not prevent new infections or inhibit active TB, but seems to help keep latent TB from turning active. BCG is routinely given to infants and small children in most of the world.

It is not used in the United States, in part because the disease is not perceived as a problem for the general population, and in part because of its limited effectiveness. Also, the vaccine can cause a positive reaction to a TB skin test, which complicates testing of children and adults.

Several labs and small biotech companies are working on variations of the BCG vaccine to make it more effective. A few are starting over, trying to fashion a new vaccine using new TB antigens that seem better able to stimulate immunity in lab animals. At least a half-dozen candidate vaccines are in various stages of development.

For decades pharmaceutical companies have shown little interest in developing a more effective vaccine, because the big markets were perceived to be in underdeveloped countries that did not have the money to make production worthwhile. The situation changed in 2000 when the Bill and Melinda Gates Foundation posted $25 million for work on new vaccines. The Gates and Rockefeller foundations are also supporting the development of new drugs through the Global Alliance for Tuberculosis Drug Development. Early in 2002 the Gates Foundation raised the stakes with more funding.

NIAID is also funding research on a TB vaccine. It sent a $2.3 million award to Corixa Corporation in Seattle, Washington, to fund preclinical and clinical testing of a new candidate vaccine produced using two pieces of Mtb proteins that are fused together. Corixa scientists tested the proteins and found good protection in animals ranging from mice to monkeys. But in a reluctant effort to cut its business losses in 2002, Corixa decided to focus its research work on cancer treatments.

In addition to preventive vaccines, researchers are looking at therapeutic vaccines—used to treat people who already have an infection. Unlike drugs that usually target the pathogen directly, therapeutic vaccines try to strengthen the immune system after an infection so it can better fight the invader naturally. This might be a more effective approach than giving yet another drug to a patient whose TB already resists other drugs. One small preliminary study of a therapeutic vaccine shows promise in people recently infected with multi-drug-resistant TB who were failing drug treatment. The vaccine was less successful in people with reactivated multi-drug-resistant disease.

Microbiologists are hampered in drug and vaccine development by a basic lack of knowledge about exactly how Mtb causes disease. There are now many different TB strains, some more dangerous than others. One avenue of investigation is to compare the genes of Mtb strains that don't cause disease or cause only minor disease with those that cause rampant TB. Hardy Kornfeld, a professor of medicine and pathology at the Boston University School of Medicine's Pulmonary Center and a frontline physician in the Boston Medical Center TB clinic, is leading this kind of work.[16] Kornfeld isolates some strains from his patients, and obtains others that scientists have grown in labs around the world for decades. He says understanding the genetic differences between strains could unveil new targets for better drugs. "If we can find virulence genes," he says, "then maybe we can knock them out."

For example, TB strains vary widely in their ability to kill a macrophage they have infected. It might seem logical that the more virulent strains would be deadlier, but the opposite is true. The more virulent strains actually keep the macrophages they infect from dying; they have realized the macrophage is a haven where they can reproduce and eventually spread to more macrophages. The last thing they want is to find themselves out in the cold. Kornfeld's work has shown that some strains even prevent T cells from instructing macrophages to commit suicide, the immune system's last-ditch effort to kill the infection by killing its own cells. "These TB bacteria have figured out the best way to exploit the human race," Kornfeld says.

Kornfeld, a slight, self-effacing man with wispy brown hair and round glasses, speaks sadly about TB's deadliness when in his lab office. But the forty-nine-year-old is also impressed by TB's defiant nature; he almost revels in it. Perhaps this is a reflection of his avocation as a motorcycle racer. The physician-scientist is a leading rider in the Loudon Road Racing Series held each summer at professional tracks around New Hampshire. He relishes taking his high-powered machine at 110 mph around a track's hard turns, his inside knee brushing the ground at extreme angles, yet all under keen control. Pushing his bike, his coordination, and his life to the limit "focuses the mind," he says.

Focusing on TB's genome offers the only real hope for scientific advancement against the disease, Kornfeld says. Microbiologists have been trying modern drugs for sixty years, and they've found few new antidotes. That said, Kornfeld notes that science has actually produced good drugs to kill TB. The bigger challenges in stopping TB, he stresses, are social and economic. "We have effective treatments," he says. "We have to get them to people, and we have to ensure treatment proceeds correctly."

Until scientists find a genetic weak spot in TB and devise drugs to attack it, or until public-health systems administer DOTS drugs to 2 billion people around the world, even countries that are relatively free of the disease will be continually threatened. To free itself of TB, the United States would have to close its borders to all outsiders. Or it would have to force every noncitizen arriving at every immigration checkpoint to undergo a TB skin test, detain them for several days till their skin reacted, and turn them away or treat them if they tested positive. But even then, they'd also have to test every American returning home from a trip to areas where TB flourishes.

Locking up illegal aliens won't necessarily help, either. Assistant health commissioner Al DeMaria notes that in Massachusetts, illegal aliens who are detained by the INS and have TB are currently treated while they are in lockups, but when they are sent back to their country, treatment likely stops. Then they try to return, and can infect new people again, maybe this time with drug-resistant TB. This is a particular problem along the U.S. border with Mexico. "There's a big move to try to improve communication between Immigration, local health departments, and the Mexican health authorities," DeMaria says, "so that someone with TB can't go back and forth without one country or the other knowing about it. The problem, of course, is that these people stay underground.

They don't come forward on their own." Nonetheless, DeMaria does think detaining people in the proper treatment setting can help reduce TB's spread, which is why his state puts patients who may be homeless or drug users in the Shattuck until they complete treatment. Patients who are admitted are counseled about substance abuse, and, if they are abusers, are given treatment. About one-quarter of the TB patients at the Shattuck are there involuntarily; the rest accept the hospitalization. DeMaria has to sign the commitment papers for each of the involuntary commitments. It's a controversial policy, but it is meant to serve both the patient and the community.

Only unerring vigilance and money will rid our country and the rest of the world of TB. No one knows that better than Michael Malone. He had already put in a long, tense morning tracking down a skittish patient before he went to Julia's office on that windy November Monday. Two days earlier, at 5:00 A.M. on Saturday, a man from Nepal had appeared at Boston City Hospital's emergency room, coughing so heavily he could hardly breathe. He said he was so sick he could no longer wash dishes at the restaurant where he worked. He had come to the United States two months earlier. He had lived with his sister in Queens in New York City for a month and a half, and now was living with his brother next to Fenway Park, the Boston Red Sox baseball stadium. The doctor did a sputum test and it was almost off the chart. The man was highly contagious and likely had been infectious since he arrived in New York. But he spoke little English and the hospital staff had trouble getting him to understand that he had TB. They worried that when they would try to explain how he would have to be held in an isolation room, he would spook and run. Malone had spent Sunday trying to communicate with him, with help from a friend who happened to work in restaurants and speak Nepalese. Meanwhile, the man was exposing people in Boston. Finally, Malone got him into the Shattuck.

Similar stories are no doubt unfolding every day in Miami, Houston, and San Francisco, in London and Copenhagen, in Tokyo and Hong Kong. Each untreated person is helping TB gain an even stronger hold over the human species, improving its means for exploiting our immune system and defying our drugs. Meanwhile, microbiologists continue to struggle in their labs, frustrated by the bacteria's poor growth and complex genome, losing confidence that mankind has the upper hand. If TB continues to resist more drugs it could kill untold millions because it is already so widely carried and so easily spread.

CHAPTER NINE

TICKING TIME BOMBS:
Chronic Diseases Combining Forces

TB is a chronic disease—one that persists in the body for years, even when treated. One of the most insidious threats in the pathogen war comes from a set of other chronic diseases that can evade our immune systems for a lifetime. The chronic infections—chief among them hepatitis C, herpes, and AIDS—have evolved diabolical tricks that allow them to destroy us silently without our even knowing they've invaded. To hold them down, our immune systems have to wage a continual battle, and if our defenders slip up even once—because they must fight off some other pathogen that's invaded, say—the chronic disease may erupt with lethal force. These diseases may even turn our own defenses against us. Hepatitis C, for example, can cause our immune system to bomb our liver ceaselessly until it is so devastated that we die of liver failure.

These "stealth" pathogens are dangerous enough in their own right, but they present an even more ominous future threat. Cohabitation of chronic pathogens with other germs increases the opportunity for them to share their secret weapons, even swap genes, which could create new superpathogens. The mixing of diseases is already causing nightmares for doctors trying to diagnose patients who have a strange combination of symptoms.

A chronic disease is also simply more likely to reemerge as a killer as you age and your immune system tires. As Michael Miller, the CDC's chief epidemiologist, says, "People are living longer, but their immune systems continue to deteriorate, making them more vulnerable." Many people may essentially become "brewers" of ever more deadly pathogens. The longer we all live, and the more susceptible our immune systems become as we age, the more we help infectious diseases defeat us.

All of these dangers stem from the schemes stealthy pathogens have learned for evading our immune system. One of the most cunning foes is hepatitis C. Once it has infected you, it mutates repeatedly, allowing it to stay one step ahead of your immune system. Just as your T cells figure out how to attack the mutated version, the virus mutates again, rendering useless the weapons your cells just devised. When B and T cells regroup and recalculate, the virus mutates yet again, and so on, so that it is able to persist in its attack on your body for decades, damaging your liver more and more. While some people infected with hepatitis C are cured by a tough regimen of drugs, in others the medications do little and they can die of liver failure.

The term *hepatitis* refers to a general inflammation of the liver, which can occur when liver cells die from prolonged stress as they try to filter excessive toxins of one kind or another from the blood. Alcoholics suffer from this inflammation, for example, because they are constantly forcing their livers to distill poisonous alcohol from their blood. Once damage to liver cells begins, it can progress to cirrhosis—the formation of hardened fibers—which can ultimately cause the liver to stop working, leading to death.

By the 1950s, scientists realized that there also was a virus that could cause hepatitis. They called this condition infectious hepatitis. There seemed to be two strains, which they named A and B. Based on patient cases, doctors thought people contracted the virus from food or water contaminated with fecal matter. That turned out to be true for hepatitis A. The symptoms of this, the least deadly of the strains, are fever, fatigue, nausea, and jaundice. People with infection often develop dark brown urine, pale stools, and a yellowish tint to the whites of the eyes fifteen to fifty days after they are exposed. Most people's immune systems clear the infection in a week or two, though the CDC estimates that 100 Americans die annually from stubborn cases.

Hepatitis A is most often spread when people do not wash their hands after using the toilet or changing a diaper. Outbreaks typically occur when an infected person with poor hygiene handles food for the public. Children in day-care centers tend to spread the disease, too, by putting their soiled hands in their mouths. Hepatitis A is also transmitted through bodily fluids during sex, but not typically through blood. The disease is so readily spread that a full one-third of all Americans are infected with hepatitis A at some time during their lives. There is no treatment, but there is a vaccine, and if you get an injection of immune globulin soon after exposure, your chances of clearing the infection without showing disease symptoms are greater.

Hepatitis B is more deadly.[1] Two billion people worldwide—one-third of the earth's population—have at some point been infected with hepatitis B. The immune system can usually clear the infection, but in some people it does not completely rid the body of the invader, which then settles in and slowly ruins the liver. In the United States, hepatitis B causes up to 19,000 hospitalizations, 32,000 chronic infections, and 6,000 deaths a year. The disease is most often spread through sex with an infected partner, but it can also be transmitted through shared needles, while undergoing dialysis on a contaminated machine, through a transfusion of tainted blood, or through any other contact with infected blood. The symptoms are the same as those for hepatitis A. An effective vaccine has been available for some time, and in the 1990s many school districts added the series of three shots to the list of vaccinations required for enrollment. There is no cure.

By the mid-1970s, researchers studying the strains felt that they understood the disease and developed a screening process for hepatitis A and B. But soon afterward, a troubling new discovery was made. Harvey Alter at the National Institutes of Health sent out warnings that yet a third type of hepatitis was getting past blood tests and attacking the livers of people who had received blood transfusions.[2] He called the disease "non-A non-B" hepatitis, and said it was very different from its predecessors.

For more than a decade this stealthy virus eluded researchers' best efforts to define how it differed from hepatitis A and B. Not until 1989, after seven years of intensive work by Michael

Houghton at the Chiron Corporation in California, was the virus finally isolated. Houghton called the new strain hepatitis C, and he began to make the antibody needed for a blood test that could detect the pathogen.[3]

This new strain turned out to be a good deal more vexing than either A or B. Its symptoms are similar, but it attacks the liver much more aggressively and with much longer-lasting determination. Today 2.7 million Americans alone have a chronic hepatitis C infection, and 10,000 die from it each year. There are currently about 25,000 new cases annually, 80 percent of which will become chronic. There is no vaccine, and drugs offer only limited help, while often causing significant side effects.

Hepatitis viruses all infect cells in the liver. Scientists think they have pieced together a cogent picture of the organisms' tactics. When helper T cells recognize the hepatitis C antigens, they turn on a class of T cells specialized to battle these kinds of intruders—the killer or cytotoxic T lymphocyte, or CTL. The CTLs multiply and head out through the body. When they find the intruders in the liver, they attack the liver cells that have viruses inside them. If the CTLs can clear the virus without killing too many liver cells, the liver will be temporarily damaged, but will recover.

Once the virus is cleared, the majority of the CTL cells commit suicide so they don't continue to attack the liver unnecessarily. This is a normal process called activation-induced cell death, or apoptosis. Something like 90 percent of the cells die, leaving a band of battle-hardened CTLs to guard the body for years against a return of the infectious agent.[4] If you are infected again with a similar-looking hepatitis C virus, the guards will remultiply and fight the infection.

However, if the first batch of killer CTL does not destroy all infected cells, a chronic infection can become established. The surviving hepatitis C invaders infect more liver cells and also begin to mutate their form. The bone marrow must manufacture more and more waves of CTL, which persist in bombing the liver for weeks, months, or years. The constant cell death leads to scarring, which prevents good blood flow throughout the liver, leading to more cell death. As more of the liver is turned into blackened waste fibers, or develops cirrhosis, the organ swells, yet loses its capacity to filter the body's blood. Toxins accumulate in the body and the victim can die of toxic shock, or the liver can shut down, causing death by organ failure. The process is slow but steady, taking an average of twenty to thirty years in people with progressive disease.

It took years for scientists to understand why hepatitis C was so lethal. When they finally discovered the virus's mutation trick, they were surprised at its evil simplicity and speed, notes Miriam Alter, chief of epidemiology at the CDC's division of viral hepatitis in Atlanta (and no relation to Harvey Alter at NIH, who sounded the first alarms about hepatitis C).

At any given moment in an infected person, millions of hepatitis C particles are infecting millions of liver cells. But rather than just fooling the cells into creating exact duplicates of the virus particles, some of the hepatitis C viruses alter a bit of their RNA genome. When the liver cell duplicates this mutated particle, it actually creates new virus particles that have slightly different antigens. Therefore, Alter explains, the CTL designed to recognize and destroy the previous generation of antigens will not clasp onto the new ones, so the new virus particles avoid being attacked. The bone marrow must produce an updated CTL. But each time a new generation of hepatitis C viruses invades new host cells, some portion of the virus makes itself over, and the CTLs never catch up. The progression is similar to what goes on with the flu virus and its antigenic drift, but at a much more furious frequency, producing more than 10 trillion copies a day. A new quasi-species can develop in only a few hours. "The hepatitis infection can create an endless cycle of new faces, for years," Alter says.

Hepatitis C is transmitted from person to person through contact with contaminated blood, and in rare cases through sexual intercourse. Soon after Houghton identified hepatitis C in 1989, the American Red Cross developed a test to screen blood for the virus. Hospitals and blood centers nationwide quickly put the test in place for blood donations and blood set aside for transfusions and surgery, to stop the flow of contaminated blood from person to person.

Nonetheless, today the disease continues to be transmitted. Shared drug needles are a leading factor. Uncleaned tattoo needles can transmit the pathogen, too, a growing concern given the greatly increased popularity of tattoos. This threat gained national attention in July 2002 when Pamela Anderson, former *Playboy* model and *Baywatch* star, revealed on CNN's *Larry King Live* that she had the disease. She claimed she had been exposed to the virus by sharing a contaminated tattooing needle with her ex-husband, rock drummer Tommy Lee.

Accidents and incompetence continue to spread the disease as well. In October 2002, officials at Norman Regional Hospital in Norman, Oklahoma, announced that they had discovered that more than fifty people out of 300 treated at their pain clinic had tested positive for hepatitis C.[5] The infections occurred because a nurse-anesthesiologist had allegedly used the same syringe and needle to inject sedation medication into the intravenous lines of several different patients on various days. Since a person's blood can back up into IV line ports, nurses and doctors are supposed to use needles only once to avoid transmitting disease. The state's epidemiologist, who led the investigation, said the nurse's actions resulted from a misunderstanding of proper procedure, which drew quick comment from nursing professionals: If this nurse thought what he was doing was okay, how many other nurses may think the same? One answer came a month later, in November 2002, when officials at the Fremont Area Medical Center near Omaha, Nebraska, voluntarily shut down its cancer and blood clinic after ten patients were discovered to have recently contracted hepatitis C, most likely due to needle reuse there.

Although transmission is less likely through sex than through blood, intercourse can also confer the disease. And people who were exposed to the virus through blood transfusions before 1989 may only be showing symptoms now, because it can lie undetected in the body for more than a decade. In that time, infected people can pass it on to others. Today, drug abusers account for 60 percent of new cases, and sexual transmission for 15 percent, with the remainder made up of dialysis patients who pick it up from unclean machines, health-care workers who are inadvertently exposed (perhaps by pricking themselves with a contaminated needle), and infants born to women who are infected. The symptoms are the same as those for the other forms of hepatitis: jaundice, fatigue, abdominal pain, nausea, loss of appetite, diarrhea, and fever. With hepatitis C virus, a small percentage of patients exhibit hepatitis symptoms soon after infection, some show symptoms decades later, and some never show signs at all. Doctors use liver tests as well as a blood test to find incriminating antibodies that indicate infection.

The CDC estimates that almost 4 million Americans are now infected with hepatitis C. Chronic liver disease is the tenth leading cause of death among American adults, and 40 to 60 percent of it is due to hepatitis C. The infection is the leading reason for liver transplants in the United States. Alcohol use or coinfection with hepatitis B or HIV significantly increases the likelihood that the liver will deteriorate.

Despite the seemingly high numbers, North America and Western Europe actually have among the lowest disease rates. An incredible 22 percent of Egypt's people are infected, largely due to an ill-fated nationwide campaign of inoculations to control the parasite schistosoma.[6] The reuse of needles by health-care workers, aggravated by the need to give people a series of twelve to sixteen injections for full immunization, inadvertently spread hepatitis C far and wide.

In 1989, when the test for hepatitis C had been developed, hospitals, the American Red Cross, and other institutions quickly set up a system for screening blood donations to stop the flow of contaminated blood from person to person. But before that time, many health-care professionals who regularly came into contact with patients' blood were vulnerable to contracting the disease, and many are sick today.

David Nathan graduated from Harvard Medical School in 1955 and before long landed a great job as a clinician at a new 500-bed hospital at the National Cancer Institute in Bethesda, Maryland.[7] The director gave Nathan a beautiful laboratory, and Nathan quickly began studying how best to apply results of clinical trials to cure cancer patients. He then returned to the Peter Bent Brigham Hospital in Boston to start his own clinical hematology lab, to research blood disorders. Like many young physicians, Nathan had been educated never to do anything to someone in a trial that was purely for the sake of research, and not necessarily in the patient's best interest. And like some of his eager colleagues, he extended that philosophy into a personal doctrine some found risky: Never perform an experiment on a patient you wouldn't perform on yourself.

In 1967 a two-year-old girl named Kathy, from Maine, was referred to Nathan's lab. She had a rare enzyme disturbance in her red blood cells and her blood cells were dying. Kathy was very anemic and required regular blood transfusions. Her spleen was enlarged. Her physician asked Nathan if he should remove the child's spleen; the enlargement suggested the spleen might be killing her healthy red blood cells as they routinely passed through. Removing the spleen could stop the killing, but the move was a big risk. Young children especially need the spleen to battle pneumococcus and other nasty organisms. Without the spleen they can die from overwhelming infection.

Nathan planned an experiment to determine whether or not Kathy's spleen was at fault. He would inject some of Kathy's blood into two people whose spleens had been removed. If her red cells still died in the absence of a spleen, then Nathan could conclude that they had an intrinsic defect and were dying on their own. Her spleen might be fine. He would also inject Kathy's cells into two people who had healthy spleens, to see if her cells survived. If not, that would indicate a problem with Kathy's spleen. Those two people would be Nathan himself and one of his lab technicians.

After several weeks, tests showed Kathy's cells were surviving nicely in the people without spleens but not in Nathan and his technician. So it seemed the girl's spleen was the problem. Her surgeon removed it, and although she wasn't completely cured afterward, she was much healthier, and no longer needed blood transfusions. "It was a good experiment," Nathan recalls. "We helped the girl. We wrote up a paper about it, and the *New England Journal of Medicine* published it in 1968. But there was a problem. Several years later the technician came down with hepatitis."

The technician sued Nathan, and a trial dragged on for five years. Even though the practice of medical professionals participating in patient experiments was common in those days, Nathan eventually settled for $25,000—the limit of his malpractice insurance. By that time the technician's immune system had actually cleared the disease. Nathan wrung his hands. *My God,* he thought. *Self-experiments, experiments on volunteers. We have to find a better way.*

Nathan tried to put the episode behind him. In 1985 he became physician in chief at Children's Hospital in Boston. It was a big responsibility, fraught with stress. He developed mild hypertension. His own internist put him on a drug called an ace inhibitor, which reduces hypertension but can lower the white-blood-cell count in a small number of patients. Early in 1990, after a regular checkup, Nathan realized his doctor had not told him about his most recent test results. So he followed the advice he gave his own patients: If you don't hear back about test results, call the doctor! All sorts of slipups can occur. Nathan phoned his physician's office, and an assistant told Nathan his white count was way down. Nathan, a senior hematologist, was furious. "How can he see my count fall and leave me on a drug that further reduces my white blood cells?! That will kill me!" Nathan asked about his platelet count. It was down, too. He got scared. "I thought I was heading toward bone marrow failure."

Nathan immediately dropped the doctor and found another at the Brigham. After reviewing Nathan's medical records, the new physician took him off the ace inhibitor and substituted other drugs. The doctor also said, "One odd thing I notice in the reports is that your liver enzymes aren't quite normal. They haven't been for several years, and they're getting a little worse. What's going on?"

Nathan thought for a moment that somehow his unusual liver results might be related to the five times he had allowed doctors to extract his bone marrow to use as a control against patients in experiments. But he and his doctor knew that was just paranoia. Hepatitis was the likely villain. Together, the doctor and doctor-as-patient ran the tests; sure enough, Nathan had hepatitis C.

Where had he gotten it from? He couldn't say for sure, but he was immediately convinced he had gotten it from Kathy's blood transfusion. His lab technician had, so why not he himself? More than twenty years after volunteering his body to help a little girl, he was suffering the consequences of his good deed.

The story of Nathan's experience with treatment is a cautionary tale about just how insidious a chronic infectious disease like hepatitis C can be. They are ticking time bombs in millions of people. And what happened to Nathan after his diagnosis shows just how hard hepatitis C is to treat. By 1990 doctors had begun to put hepatitis C patients on interferon—a pharmaceutical form of the same cytokine the immune system produces naturally. Theoretically, extra interferon would help the CTL immune cells kill off the virus. But in practice it helped only 10 percent of patients. "You take it for two years, and you're miserable with heavy flulike symptoms for every minute," Nathan says. "I wasn't going to do that. I figured the Good Lord would protect me." Luckily, he wasn't feeling much ill effect from the infection.

In 1995 Nathan became president of the Dana Farber Cancer Institute, the preeminent U.S. cancer research center. His internist, who remained worried that the early fibrosis might progress, tracked advances in hepatitis C treatment and eventually referred Nathan to a hematologist at the Brigham. New trials showed that if hepatitis C patients took interferon plus ribavirin, a drug that had just been approved, there was a 25-percent success rate—still not great odds, but better than 10 percent. Ribavirin is an all-purpose antiviral drug that also interferes with a virus's ability to replicate. Patients had to take the pair of drugs for six months to a year, and there were significant side effects. But there just was no other treatment.

Nathan, growing more worried himself as the medical community issued stronger evidence of the disease's deadliness, started the regimen in 1998. He took ribavirin pills daily, and three times a week he had to get interferon injections. The symptoms were a nightmare. Each injection left his thin frame exhausted. "When you have any infection, your immune system mounts a counterattack against the organism," he explains. "It releases all sorts of cytokines, which cause the symptoms. I felt like I had the flu all day, and badly, but without the fever. Meanwhile, ribavirin slices up red blood cells, which makes you anemic and jaundiced." Nathan's skin turned yellow. The combination of drugs also frequently leads to depression, and the patient is fighting desperate fatigue all the time.

Nathan struggled through every hour of every day, trying somehow to run Dana Farber in the midst of his battle. He would collapse on the couch in his living room, by the baby grand piano. His only solace from suffering came when he played music recorded by his daughter, a professional flutist. The music was Nathan's haven.

Spurred on by hope, Nathan asked his internist to take frequent blood samples to measure the load of hepatitis virus in his blood. He knew that if the drugs could help kill the virus quickly, he had a good chance for recovery. But if the virus died off slowly, it would have time to develop drug resistance, and he would not recover. His viral load dropped like a shot. In three weeks he was almost virus-free. Nathan knew he would be a survivor.

"Almost virus-free" was a key phrase, however. Nathan had to continue the drugs for much longer. If even a few hepatitis C viruses remained hidden in his body, they could flair up again later. After all, it had hibernated inside him for two decades before erupting. It was a time bomb that finally went off. But six months into the treatment, Nathan developed an allergic reaction to the ribavirin. While he was visiting Tuscany in the spring of 1999 with his wife and some close friends, Nathan's skin began to itch so badly he had to cut the vacation short and return to the United States. Doctors tried to treat his skin with steroids and medicated baths, but lesions would open and bleed. "I was an unbelievable mess. My skin was so awful I couldn't even sleep." Finally his internist tempered justice with mercy and asked Nathan if he wanted to go off the ribavirin and interferon. Nathan said yes. The disease could gradually return, but Nathan chose to again place his faith in God, rather than continue to suffer a living hell.

To this day, Nathan's liver chemistry has remained normal. He could relapse, but the chances now seem slim; if the disease was going to resurge it probably would have by now.

Nathan has no idea what happened to the two other adults whom he also injected with young Kathy's blood. They were much older than he was, so he surmises they are long gone, whether they suffered from hepatitis or not. And despite having settled a lawsuit, he still can't even say for certain whether he or the technician actually got their infection from Kathy, whom he had no reason to keep track of. Nathan and his technician handled blood as dangerously as everyone else in the 1960s and 1970s. He never wore gloves, and had chapped skin and abrasions on his fingers all the time, which gave the raw blood easy access into his veins. He washed his hands, but like his contemporaries he was casual about it. Today many hematologists and oncologists who were working before blood screening became standard practice in 1990 have hepatitis, and very few of them will ever be able to pinpoint exactly when and how they got it, or how long they may have been capable of spreading the disease to others. To handle blood without gloves today would be anathema, and to transfuse the blood of a little girl who had already had other transfusions into any volunteer would never be allowed. "In those days," Nathan says, "we just didn't know."

Countless people outside the medical profession also picked up hepatitis C through blood transfusions before the blood-screening tests were finally in place nationwide in 1992. But because the virus can have a long latency period, cases of hepatitis C rose exponentially through the early 1990s. The human immune system clears the pathogen in only about 20 percent of people. For the rest, the virus multiplies, hides out, and will tick away inside them for the rest of their lives, perhaps surfacing, perhaps not.

Once blood screening was standard, and as public-health officials educated people about the danger of sharing drug needles and having unprotected sex, the number of new infections per year dropped to 25,000 in 2001—still a huge number. Worldwide, the number of new cases is daunting. More than 170 million people now have hepatitis C, and 3 to 4 million more are infected each year. The World Health Organization now says that by 2006, hepatitis may kill more people annually than AIDS.

Because hepatitis C was only discovered in 1989 and the symptoms can take ten to twenty years to develop, another decade may go by before we know just how many people will die from it. It will be telling to see whether the percentage of those in whom the disease progresses to cirrhosis or cancer declines or increases during that time. Nearly 3 million Americans are potentially on the road to cirrhosis, liver failure, and liver cancer, not counting legions of undiagnosed people in prisons and on the street where the disease runs rampant due to unsafe sex and drug use. Medical statisticians say that 10 to 20 percent of people with chronic hepatitis C develop cirrhosis, and 10 to 20 percent of those with cirrhosis develop liver cancer.

Virologists have recently identified two additional types of the hepatitis viruses. Hepatitis D is essentially a defective virus that cannot reproduce on its own; it can only infect and multiply in people who are already infected with hepatitis B, by using an extra hepatitis B surface protein to complete itself. But when it appears in conjunction with hepatitis B, it makes the illness worse. Hepatitis E is similar to A, and is rising mostly in underdeveloped countries, but unlike hepatitis A, a shot of immune globulin does nothing to control infection. Like hepatitis A, it is readily cleared by the immune system. The illness it causes is somewhat more severe than hepatitis A, and is especially dangerous for pregnant women, killing 20 to 30 percent of those that are infected. What these strains portend for the future is a mystery.

All told, one in twenty Americans will contract some form of hepatitis. Despite widespread education about the risks of drug abuse and unsafe sex, the determined hepatitis C virus continues to flare up everywhere. In June 2001, when New York City assistant health commissioner Marci Layton was in her office preparing her agency for the summer's oncoming invasion of West Nile virus, two staffers suddenly appeared to tell her that the boss had just called an emergency meeting. Doctors had reported a cluster of eight patients with hepatitis C. They traced the outbreak to an endoscopy center in Brooklyn. Layton spent the next morning at the center and, based on what she saw, guessed the pathogen was moving from patient to patient through unclean anesthesia equipment. The likely scenario: misuse of a contaminated syringe, like that found in the Oklahoma case.

Layton's office quickly had to track down all of the center's recent patients and test them for hepatitis C, as well as hepatitis B and HIV. If contaminated equipment was spreading one disease, it could be spreading the others, too; all it takes is one patient infected with each disease. Layton's crew had to contact each patient's spouse or sexual partners. Six months later they would have to retest people who had tested clean, too, because it can take several months for a newly exposed person's immune system to make antibodies to the virus, which are what appear on the tests. The health department offered free testing and free transportation to testing sites for everyone who had been treated or examined at that center. Layton set up a hotline that physicians and patients could call to reach a staffer able to answer questions. Layton had learned the importance of such a hotline through her experience dealing with the West Nile outbreaks the previous two summers.

Since scientists discovered hepatitis C, they have been trying to figure out how to stop it, but they have been frustrated by a vexing problem. For some unexplainable reason, the virus simply refused to grow in laboratory cell cultures. That has made it nearly impossible for microbiologists to understand in any depth how the virus interacts with cells and reproduces, and thus how to develop a vaccine or drugs. There are also no good animal models of hepatitis C infection. Chimpanzees are the only animals other than humans that contract the virus, but chimps reproduce slowly and are expensive and difficult to handle in the lab. In 2001, however, Norman Kneteman at the University of Alberta in Canada found a possible way around the problem, when his team engineered mice that accepted mammalian liver cells that grow as nodules within their own livers.[8] The team has since introduced the virus into the mice, and has shown that interferon can help mitigate the disease in some cases. But this only brings scientists to what they already know occurs in humans. Kneteman's group plans to investigate ribavirin next, to continue to see if hepatitis and antiviral drugs in mice mimic what happens in people. If so, they will have more confidence that the mouse model will be useful in predicting what new drugs may do in people. Perhaps then they can advance into uncharted territory.

The other avenue is to map hepatitis C's genome and try to find weak spots. Late in 2001, two companies, Innogenetics in Belgium and Chiron in the United States (which originally discovered hepatitis C), reported limited success in early human trials of a vaccine with genetically engineered antigens that might induce recognition of some of the virus's coat proteins. The vaccine is being used in people who are already infected, in the hope of tipping the balance against the killer cells. The vaccine would not prevent infection, but could help the immune system keep the disease in check so that cirrhosis or cancer would be less likely to develop.

The best treatment available at this time for hepatitis C is almost exactly what Nathan was taking: a combination of interferon and ribavirin.[9] The only improvement has been to add a carbohydrate tag to the interferon so it circulates in the blood longer. The new drug is called PEG-interferon. But even though the combination therapy has been improved since Nathan's treatment, it still helps fewer than half of hepatitis C's victims.

In China, where hepatitis is more widespread than in the West, the two-drug combination is now ineffective in 60 percent of patients, because of drug resistance. This situation has quickly arisen from one of the virus's most frightening qualities: It can mutate so fast that it can develop newfound drug resistance *after* it has already infected you.

Chronic infections can silently destroy us over time without our knowing it. And in a growing number of cases our immune cells contribute to the problem, not the cure. Recall that as vCJD eats away at the brain, the microglial immune cells try to kill the misfolding prions with toxins, poisoning healthy brain cells in the process. Hepatitis C sometimes tricks our immune system into ceaselessly bombing our liver until it gives out, killing us. In both cases a stealthy pathogen quietly sneaks into our bodies and turns our immune system against us.

The hepatitis C virus has ways of preventing your immune system from defeating it. Tuberculosis knows how to sneak past your immune defenders to begin with. What if a pathogen knew both sets of tricks? There is such an organism. And there's a 90-percent chance it's inside your body right now. It's herpes.

Herpes? That "sexually transmitted disease" that gives people occasional sores? Yes, to an extent, but it's not that simple.

"Herpes" is actually a family of incredibly diverse viruses. Herpes simplex 1 generally causes the familiar cold sores many of us get in our mouths at some point.[10] But it can also sometimes cause genital sores. Much worse, it can cause a form of fatal encephalitis, cases of which have doubled to 2,000 in the last ten years, killing 70 percent of victims who go untreated.[11] It can also cause herpetic keratinitis, which is a leading cause of blindness in developed countries. Herpes simplex 2 causes genital sores, but can also cause meningitis and is responsible for more than 50 percent of fetal deaths worldwide, infecting unborn babies through their mothers' bloodstreams. Human Herpes 3, known as the varicella-zoster virus, causes chicken pox and can reactivate to cause shingles in adults. All these viruses tend to establish chronic infections in your nerve cells.

Human Herpes 4, also called Epstein-Barr virus, actually infects B cells, causing infectious mononucleosis—the bane of teenagers—but it can also advance to cancer of the pharynx and to Burkitt's lymphoma, a cancer caused by aberrant B cells, and it, too, can cause encephalitis. Human Herpes 5, cytomegalovirus (CMV), infects macrophages, and can cause jaundice, brain damage, and death. Human Herpes 6 is associated with a mild infant infection, yet may contribute to multiple sclerosis. Human Herpes 7, together with 6, can facilitate infections in transplant patients. Human Herpes 8 lies dormant in most people, but can suddenly transform normal human cells into the cancerous Kaposi's sarcoma or B cell lymphoma—and recent research shows that people can pass it along in saliva in rare cases, although it normally is acquired through sexual contact.

Herpes was first recorded in ancient Greece, and has been haunting us ever since. Incredibly, 90 percent of all adults—in every developed and underdeveloped country—have at least one type of herpes virus already inside them. That most likely means you. And once you contract any herpes virus you have it for life. Luckily, herpes remains latent in most people for most of their lives. But it can activate at any time, usually when your immune system has let down its guard, making it easier for the herpes virus to evade your protective mechanisms.

Currently doctors treat herpes infections with a family of several drugs, but there is no cure. Drug treatment primarily reduces the length of reactivation episodes, such as an outbreak of cold sores by a day or two. Vaccines have proved ineffective so far.

Humans are the only source of herpes infections—not tainted food or water, not mosquitoes or other insects or any other animals. The virus rides in people's bodily fluids—saliva, blood, semen, and vaginal fluid. All you have to do is rub your eyes, kiss someone, share a toothbrush, brush up against someone else's infected skin, or have sex with an infected partner, and herpes has an easy pathway into your body.

When a herpes simplex virus first sneaks in, it worms its way into your nerve cells and works its way up tiny nerve fibers to the body's main nerve lines in places such as the jaw and hips. Nerves provide a clever haven because immune cells typically don't inspect them looking for intruders.

Once the virus has infiltrated the body's nerve lines, it lies low and waits. Usually, when a virus tunnels inside a healthy cell, it makes proteins that the host cell will duplicate to create more virus particles. But herpes ekes out just the few proteins it needs to sustain its genome. Furthermore, if the cell's antigen-presenting molecules—the MHC inspector molecules—do find the proteins, the proteins contain very few antigens. The MHCs therefore have little to present at the cell's surface, and T cells do not become aware that an infectious virus has invaded.[12]

Herpes can also make a protein that binds to the MHC molecule and prevents it from grasping even what few antigens there are. The Epstein-Barr versions of herpes can actually stop the cell from making MHC molecules so there are fewer inspectors to transport antigens to the cell surface.

Every now and then herpes sends out a representative to scout the scene. The viruses act a bit like bandits in the old West, who would establish a hideout in the wilderness, occasionally surfacing to rob a bank or train.[13] When the scout sees that immune defenses are flagging because you're tired or stressed, or are fighting off another virus or bacteria, it signals the other herpes virus particles that it's time to stage a raid. They move back down the nerve to the original area of infection and break out. The EBV and CMV herpes viruses that infect immune cells lie low in similar fashion, then break out within the immune system when it is down.

Once herpes erupts, the body's immune system tries to battle it back down. But herpes cuts the attack short by retreating. The body then thinks the invader has been beaten, and stops fighting—leaving the door wide open for the herpes to come roaring back the next time it senses a weak moment.

Despite herpes' best efforts to hide, MHCs may eventually get wind of some herpes antigens and bring them to a cell's surface. T cells will detect the presence of intruders and begin to make cytokines, including interferon, the cytokine used in drug form to combat hepatitis. In these cases, herpes works to limit the consequences. Like hepatitis C, herpes simplex 1 and 2 attempt to prevent the interferon at the cell's surface from delivering the message to the cell's internal command center to activate genes that would halt further virus production.[14] The Epstein-Barr virus, in contrast, tries to trick blood cells into releasing a different cytokine that acts like a brake on the immune system, and for good measure it makes more of the cytokine itself from a gene stolen long ago from an infected person and now shared by all known Epstein-Barr viruses.[15] This cytokine is called interleukin 10, and it keeps T cells from making more interferon. It also encourages the T cells to manufacture useless antibodies rather than mobilize helper T cells and CTL to fight the infection.

If all this fails, herpes is perfectly capable of engaging in hand-to-hand combat. When Epstein-Barr realizes T cells have arrived to investigate, it fools the infected B cells into commanding the T cells to self-destruct.[16] Usually the T cells put up enough of a fight to keep the virus under control, even if they can't eliminate it. But in relatively rare cases the misguided B cells go haywire and turn into cancer cells, causing Burkitt's lymphoma. Most Burkitt's cases are spread across central Africa, which researchers have realized coincides with the malaria belt. It turns out the confluence is not accidental. Malaria weakens the T cells that would normally control the infected B cells, making Burkitt's more likely.

When translated and boiled down for a short explanation in a book, herpes' operating mechanisms seem straightforward. But they are not. Indeed, the explanations scientists have generated for how herpes, hepatitis C, and TB—and most pathogens—work have gaping holes. Much is inferred from how they behave in lab cultures, which may or may not be how they behave inside our bodies.

One of the first direct, visual observations of the immune system in action is currently in progress, under the direction of Hidde Ploegh, professor of immunopathology at Harvard Medical School in Boston.[17] The tall Dutchman and his twenty-five lab people, including a noticeable contingent of students from Ploegh's native Netherlands, are using black mice to show how a macrophage in the skin actually identifies an intruder. They have altered a mouse gene so that when it makes proteins that control the MHC inspector molecule, the proteins fluoresce. Using a high-powered microscope and special cameras, they have created a video that shows what an actual MHC molecule does when its macrophage grasps a particle.

Ploegh says herpes is so advanced it understands the human immune system better than we do, so he studies how the virus evades our defenses to learn about us. Herpes may cause great damage to our bodies, but it may be a handy tool with respect to the development of our understanding of how our bodies fight disease. "We really know so little about how immune cells and pathogens interact," Ploegh says. "We don't even really know how many types of phagocytic cells there are. We are trying to define the actual rules of engagement. Only then can we find the best ways to interfere."

If hepatitis C is ingenious and herpes is the ultimate stealth organism, there is another virus that has developed an even more advanced capability to ensure its successful life in our systems: HIV.[18] It doesn't worry about stealing resistance genes from other pathogens—it easily develops them on its own. HIV is the ultimate warrior. It infects both helper T cells and macrophages.

HIV is a master at resisting drugs. Like hepatitis C, it has a very high mutation rate. It can change the looks of its enzymes so drugs don't latch on. That's why doctors put HIV patients on a cocktail of three or more compounds—so HIV must simultaneously come up with three or more drug-resistant mutations at once. Unfortunately, the drug combinations never actually kill HIV; they only stop the pathogen from infecting more cells and temporarily stop the killing of T cells.

Unlike hepatitis C, which sometimes remains in remission after drugs are discontinued, generally, the minute an HIV patient stops taking his drugs the virus resumes its attack. HIV patients lack the option of those with hepatitis C of not taking drugs in hopes that their immune system will clear the disease. They have no choice but to endure the often difficult side effects of drugs, ranging from severe anemia and loss of white blood cells to liver and muscle toxicity.

Unlike the cat-and-mouse games that herpes plays with the immune system, the HIV virus just plain attacks the heart of the immune system directly, killing off the very helper T cells that would keep HIV in check if given the chance. For you to contract HIV, an infected person's semen, vaginal fluid, or blood must enter your body. Even then, transmission is infrequent. Nevertheless, HIV has spread globally in less than twenty-five years. It has infected more than 50 million people and killed more than 20 million of them.

Your immune system may valiantly battle back HIV for several years, but the invader is relentless. Sooner or later the organism gets the upper hand, and it becomes more and more destructive as it gains advantage. Most HIV virus particles infect helper T cells and destroy them from within. Some send internal messages that tell the cell to kill itself. Others trick killer T cells that brush by into thinking that an innocent helper T cell is an enemy, and the killer T kills the uninfected helper T.

The T cells do manage to put up a fight. After all, it takes a person infected with HIV an average of eight to ten years to develop full-blown AIDS. AIDS is defined in people who are losing the battle against HIV when they reach a very low number of helper T cells in their blood, or when they develop certain conditions such as *Pneumocystis carinii* pneumonia infections or cancers such as Kaposi's sarcoma that point to a severely weakened immune system. At first, helper T cells hold steady at about two-thirds the normal number as killer T cells fight the virus, but the helper T cells then start to drop each year until they reach about 20 percent of normal. This is when AIDS is usually diagnosed.

At such low concentrations the T cells lose their ability as a group to battle infections. To make matters worse, the other immune troops that look to the helper T cells for guidance no longer get enough useful information from them; B cells make fewer antibodies against new infections, and phagocytic white blood cells become less effective at killing invaders. The victim becomes prey to common infections, as well as to all sorts of unusual infections that rarely arise in healthy people because their immune system is easily able to contain them. That is what is behind the name "acquired immune deficiency syndrome" and the term "opportunistic infections." When you are fighting AIDS, you have almost no immune system left to fight anything else; every "routine" infection poses a dire threat.

Researchers have created several candidate vaccines, but early trials have been far from promising. HIV's ability to change rapidly is a major problem, rendering a vaccine ineffective. Vaccines also prove elusive because, as is so often the case in the world of infectious diseases, scientists' understanding of HIV's inner workings is educated guesswork. Deenan Pillay, a leading HIV researcher, acknowledges that inadequate level of knowledge freely.[19] Born in Britain to parents from South Africa, the dark-haired Pillay toils away at Birmingham University in England, a proper fortress of marble floors and columns teeming with English and Indian students. What scientists need to know most about HIV is how it manages to mutate so quickly that it can develop resistance to a vaccine or killer drug after it has already settled in a person's body.

HIV can produce a billion viruses a day, constantly serving up mutations that defy drugs.[20] "If you treat an HIV patient with only one drug, resistance can develop in a week," Pillay says. "A recent study in Africa showed that even a single dose of a single drug can lead to resistance." If in one day only 100 of the 1 billion new copies have a mutation, and only ten of those happen to be less affected by a given drug, those few virus particles can replicate enough to undermine treatment, with their generations of offspring killing and infecting cells at a rate fast enough to outpace the number of immune cells the drugs protect.

Pillay and others have also discovered another frightening reality. People with HIV can transmit drug-resistant strains to others. As HIV is passed around, its ability to resist more drugs and drug combinations mounts, and it could outrun the pharmaceutical industry's ability to devise, test, and distribute new compounds.

Today there are only about twenty-five drugs that work against the entire list of viruses mankind faces, and fifteen of them have been developed to fight HIV. Most of the drugs target a specific enzyme that the virus uses in replicating its genome inside the host cell. But fifteen drugs against HIV are not a lot. We may soon see patients with resistance to all of them if we don't come up with alternative treatments quickly. The clock is ticking. Fortunately the strategy of using three or more drugs at a time, and testing patients for drug resistance and then prescribing compounds their HIV is not resistant to, has slowed the evolution of drug resistance in HIV.

To really understand how HIV develops resistance, Pillay must study virus from victims who have only recently been identified as HIV positive, but have not yet started on drugs. Finding such patients is difficult enough; Pillay continues to develop an ever-wider network of contacts in clinics throughout his country and others who will refer patients to him, and yet few patients who are newly diagnosed are willing to delay their own treatment even an extra few days so Pillay can get the information and blood samples he needs. When he does succeed, the ensuing genetics work is amazingly complicated. Pillay and colleagues around the world have identified 250 mutations in the HIV genome that can trigger drug resistance.

Just preparing a research sample is taxing. Pillay must obtain a person's blood, spin it down in a centrifuge to extract the plasma (which contains HIV), and use a lab technique known as PCR to detect and amplify certain stretches of its genome to find telltale sequences of nucleic acids that indicate potential resistance genes. There are typically about 1,000 nucleic acids to consider.

At this stage Pillay must submit the long, suspect sequences over the Web to one of several commercial services that will compare them against a master HIV genome database run by Los Alamos National Laboratory in New Mexico. The computer-intensive comparison will suggest whether the patient's sequences match certain sequences that scientists have shown create resistance to a particular drug. Unfortunately, Pillay says, thus far all the services base their matching software on one HIV subtype known as HIV-B, because it prevails in the West. But virologists have already identified eight subtypes, A through H. While HIV-B accounts for 95 percent of cases in North America, it accounts for only 70 percent in Britain, because the latter's population includes many more people who have immigrated for generations from places such as India and South Africa, where other subtypes predominate. Subtypes A, C, and D, for example, prevail in Africa, and E prevails in places such as Thailand. Infection in Nigeria, Ethiopia, India, China, and Russia is expected to explode in the next several years, with 80 to 110 million cases predicted in those countries by 2010.

Clearly the best way to prevent the continuing AIDS epidemic is to educate people about how to avoid risky behavior that can allow an infected person to transmit the virus in the first place. The United States supports an ongoing media campaign encouraging people to reduce their number of sexual partners, to use condoms routinely, and to discard used needles. But many of the highest-risk countries are too poor to mount such efforts, which is why various international agencies are now targeting portions of their funding for these purposes. Yet even with all of our knowledge—and fear—the virus continues to find new victims here. Amazingly, even in 2002, 77 percent of gay men ages fifteen to twenty-nine who were randomly chosen for a study and turned out to be HIV positive were previously unaware that they were infected.[21] Imagine how the killer virus could spread in countries such as China, India, and Russia, where knowledge about prevention is not nearly as widespread.

Of all the chronic diseases, including TB, hepatitis C, and herpes, HIV presents the most damning case of how destruction of a victim's immune system makes him more vulnerable to other infections. Its spread has also made scientists more worried than ever about our growing vulnerability to dying from multiple infections. Next winter you might fight off an influenza attack. In the spring you might clear a hepatitis C infection. In the summer, with heavy drugs, you might beat down a TB invasion, too. But in the real world your immune system doesn't have the luxury of confronting one disease at a time. Battling more than one pathogen divides your immune system's forces, and it's much harder to fight a war on two fronts. Having TB or hepatitis C or HIV alone is terrible enough, but if you have latent TB and then get HIV, you are 100 times more likely to develop active TB, and you are less likely to respond to drug treatment. If you have hepatitis C and get HIV, you are much more likely to end up with severe liver disease.[22]

We hear again and again that any infectious disease, even a mild one, can be a serious threat to people with "compromised immune systems." A chronic infection such as HIV turns you into that very person. You, ordinarily healthy, can suddenly become extremely vulnerable. One lone infection could start a landslide. This fate is similar to that suffered by cancer patients who undergo chemotherapy; the treatments kill many immune cells in the effort to kill cancer cells, making it much easier for infections to prevail.

Being infected with two or more pathogens is more than doubly dangerous. Influenza leads to so many deaths mostly because it taxes the lungs heavily, allowing other viruses and bacteria that normally don't pose much of a threat to them to cause pneumonia. Most children who get measles and die are not killed by that virus; they die from other infections such as bacterial pneumonia, which erupts because their immune system is down. Fortunately, the flu and measles pass in a week or two, so you are not vulnerable for too long. But when your immune system must constantly contend with a chronic infection like HIV, it is always compromised.

Furthermore, as pathogens accumulate in your body, they have more opportunities to share their tricks—the ones that turn *E. coli* into deadly O157, that spark flu pandemics, that make staph defy drugs. If HIV handed over to another retrovirus its superior ability to beat drugs by mutating rapidly again and again, that virus might suddenly become just as unstoppable. If they didn't kill us when they first infected us, an *E. coli,* Lyme disease agent, or West Nile virus that took on the genes that allowed it to hide safely away in our bodies could explode years later into an infection run amok, like flesh-eating bacteria, or into a rapidly spreading cancer, like Burkitt's lymphoma. And what if the ubiquitous but quiet herpes absorbs immunosuppressing genes from measles? As we spread these menacing diseases, we raise the chances that more and more people become ticking time bombs. Without actually curing chronic diseases, as we live longer as a species we become more dangerous to ourselves.

Coinfections and crisscrossing pathogens are also making it incredibly difficult for doctors to figure out what is wrong with a patient, putting the victim at even greater risk from a slow diagnosis or improper treatment. Doctors in New York City regularly curse that challenge during their special weekly exercise, called Intercity Infectious Disease Grand Rounds, which has been held each Monday afternoon for thirty years.

Some thirty-five physicians attended one recent meeting at Bronx-Lebanon Hospital in the South Bronx. The doctors—black, white, Asian, Indian, Irish, British, male, female—from the city's five boroughs and surrounding counties matched wits. One doctor in a suit bragged about the scrumptious lobster dinner he had had the night before, while another, in a tattered lab coat, complained about not being able to get enough money for his clinic to keep a supply of syringes.

As each attendee arrived, the host doctors handed them one-page synopses of seven real but unnamed patients. Each listed the patient's symptoms, brief history, and lab test results—the same information your doctor would have to diagnose you. Except that these one-page synopses described the most perplexing cases to turn up recently in area hospitals. The point of the session is to help the doctors keep their diagnostic skills sharp and learn from one another about how to decipher the confusing, often hidden nuances of the complex illnesses coursing around—what to watch for as new patients arrive on their doorsteps with confusing symptoms.

When it was time to begin, the actual doctor for Patient One stood in front of the group and entertained discussion. The veterans and newbies began shouting out questions, asking if the patient had complained of a certain symptom, if he was taking a certain drug, what an X ray might have shown. They then started calling out possible diagnoses, giving their reasons. To a degree they were competing to show how clever they might be, and to a degree they were all muddling through the case together. The presiding physician wrote each guess on a blackboard as doctors called them out. Members of the group challenged one another's hunches, argued, debated, then agreed to cross off suggested diseases as each failed to hold up to their collective scrutiny. After fifteen minutes of discussion about each patient, the physician would let them in on the actual answer—if there was one. Some of the patients had been successfully diagnosed, some had not.

Two hours later, on this particular day, even this seasoned group grew alarmed at the dramatic variations of infectious diseases that had arisen in just the past few weeks. For instance, Patient One, a fifty-year-old African-American man, had been infected with HIV and hepatitis B, and ended up with a bizarre bacterial infection called *M. kansasii,* which resisted four different drugs and left him with a swollen stomach and hardly able to breathe. The doctors agreed he had developed the strange infection because the hepatitis and HIV made him susceptible. They also commented aloud that soon they would have to start putting any patient who seemed to have hepatitis B on two drugs from the start, because resistance was growing; this observation hadn't been documented in the scientific literature yet, but they were all seeing it in more and more patients.

Patient Three, a forty-one-year-old Jamaican man, had fever, chills, rash, blistered lips, intense throat pain, greenish sputum, and blurred vision. He had recently battled pneumonia. Doctors, suspecting HIV, had put him on a cocktail of four drugs, but they did nothing. The seven leading guesses on the blackboard were all diseases caused by herpes, meaning most of the doctors were thinking along the same lines. But neither the group of thirty-five physicians nor the actual doctors were ever able to determine the man's problem. He died of toxic shock.

Patient Six, a forty-two-year-old female, had been weak and vomiting for nearly three weeks. She had had the herpes virus that causes chicken pox and shingles, and was HIV-positive, though she had stopped taking her four HIV drugs when she began vomiting during a recent visit to her homeland in the Dominican Republic. When she stumbled into the New York emergency room, doctors injected her with IV fluids to rescue her dehydrated body. She had lost thirty pounds since her previous HIV clinic visit, five weeks earlier. The roomful of doctors were baffled by what caused her vomiting and extreme weight loss. "It could be listeria," one said, referring to the food-borne bacteria that can cause fatal infections. "Did you do a spinal tap?" The presiding physician said yes and gave the results. "Maybe botulism," another called out. "Did she eat canned food in the Dominican Republic?" The physician didn't know. "Lyme disease could do this," a third doctor ventured.

The blackboard list got very long. When the guesses stopped coming, the physician informed the group that the woman had died of respiratory distress. An autopsy showed that a cyst had developed in her brain. He projected an MRI image of the affected area on a screen. The blood vessels surrounding the cyst were jammed with dead macrophages that were filled with toxins. The neurons were dead. The woman's brain had swelled, jamming the area that controls her breathing, killing her. The physician revealed that in his medical team's best analysis, the cause of the brain damage was most likely some combination of the effects of chronic HIV and herpes, with the addition of a nasty, unidentified bacteria. The condition had no name. At the thought of yet another unknown, untreatable, gruesomely fatal disease, the thirty-five doctors let out a collective sigh.

Tired by the end of the session, the doctors shook their heads in common grief as they exited the room. Five of the seven patients had dire complications due to chronic diseases.

The spread of chronic infections, combined with the rise of new and reemerging bacteria and viruses, raises the odds that mankind will have to face superpathogens it may have little chance of beating. And the chronic diseases' ability to resist drugs by mutating rapidly means our pharmaceutical industry may not be able to provide much help.

Strangely, herpes could provide the greatest potential threat. It has already mastered how to pass from person to person—unlike HIV or hepatitis, which for the most part require our bad habits to spread. If herpes were to share its contagious abilities with hepatitis C or HIV or other, more aggressive viruses, the human race would be in serious trouble. We would be engaged in a deadly waiting game, walking around with an increasing number of time bombs ticking inside us since herpes viruses, hepatitis C, and HIV have already become masters at evading immune responses.

We coexist with chronic infections in a delicate balance. In recent times herpes viruses have shown signs of going erratic. People are dying from herpes encephalitis. Epstein-Barr virus fuels Burkitt's lymphoma. We have no vaccine or drug to prevent or defeat EBV, and Burkitt's responds poorly to radiation and chemotherapy.

These sorts of superpathogens might remain curious possibilities. Or they could arise from competition among disease organisms. As our bodies carry more chronic infections, they might conflict. The bacteria that cause Legionnaire's disease and the sexually transmitted chlamydia bacteria both try to disable phagocytes in similar ways. If they find themselves inside one person's body fighting for a foothold, one of them might mutate a highly destructive mechanism that defeats its competitor—and defeats us in the process.

CHAPTER TEN

WHAT WE MUST DO

The outbreak of SARS, the emergence of superpathogens, and the increase in commingling of chronic infections all drive home just how complex a fight we must wage. We are in the midst of an evolutionary war against microbes that have proven more sophisticated and multitalented than we could ever have expected. We cannot underestimate how serious the struggle is. It's obvious that our bodies alone are unable to outwit many pathogens. Now our drugs are failing as well. If we don't step up our efforts to strengthen our defenses, the microbes might actually win in the long run. Unless we mobilize our resources more effectively, the risk is too high that there will be widespread suffering as well as social and economic disruption.

Is this outcome just an idle worry? The CIA doesn't think so. It had become so concerned by the pathogen threat by 2000 that it produced a report that sternly stressed that today's pathogens could seriously weaken the American people, their military, even their way of life.[1] Rising numbers of outbreaks might actually thin the ranks of the military enough, the report warned, that we might not be able to defend our homeland adequately. At the same time the country's productivity could wither with a sickened citizenry, and if the nation was racked by disease, the cost of health care could further pillage the economy. The October 2001 anthrax scare also made it clear that a national health crisis could undermine the very fabric of daily life, as fear and uncertainty grip the nation, and SARS inflicted significant disruptions in several countries.

The few people still among us who lived through the terrible smallpox, TB, and polio scourges of the early 1900s will tell you emphatically about how dramatically daily life was affected. The medical and scientific communities certainly know more now than they did then, but if superpathogens continue on their steep evolutionary curve, we may find ourselves facing scourges every bit as devastating.

The sobering fact is that to a large extent we have inflicted this situation on ourselves. Our relentless alteration of the earth's environment is contributing to the microbes' evolution. We began spraying the insecticide DDT in the 1940s to kill mosquitoes that spread malaria, but by the 1960s mosquitoes had evolved to resist it, and malaria rages unabated today. When penicillin was first commercialized it killed all staph infections, but now more than 90 percent of them resist it. What exactly is "resistance"? It is evolution—at a blistering pace. A pathogen hurriedly mutates new genes or acquires ready-made ones that give it new properties. By attacking organisms with chemicals and drugs, we've forced the bugs to evolve, and may have accelerated the rates at which they are changing. The HIV virus, remember, can create drug resistance in just one week after being exposed to just one dose of a drug.

By pushing pathogens, we may not be just speeding up their evolution; we may also be altering the methods by which they evolve. "We certainly don't understand enough about the pressures we are exerting on them," says Hidde Ploegh at Harvard. Without question, they can change faster than we can. And despite how impressed we are with our own magnificent evolution, pathogens couldn't care less. "There is no reason to believe humans will exist forever and not become extinct," Ploegh says. "It makes no sense biologically."

Boston University's TB expert Hardy Kornfeld agrees. "We understand, for example, how TB works, but it does not necessarily have to stay that way. It could develop a new strategy that is even more destructive."

Pathogens may be altering human evolution as well. When a human population is decimated by an infection like the plague or HIV, people whose genes give them a survival advantage do better, and the human species advances in a slightly different direction. But that direction may not necessarily make us healthier; it could make us more vulnerable. For example, in 2001 scientists at the University of Bonn in Germany found that certain people had a genetic mutation that helped fend off HIV infection, yet that same mutation made them more susceptible to hepatitis C.

At the same time, we may actually be tipping the evolutionary scales in the pathogens' favor by weakening our own immune systems. Young children who grow up in ultraclean homes may have less robust immune systems because they are not challenged enough. Adults who take antibiotics at the first sign of a cough or runny nose end up killing mildly threatening bacteria that happen to ride along with the germs that are causing the cold or flu, which prevents the immune system from being fully exposed to our enemies. "In this way," Japan's drug-resistance pioneer Keiichi Hiramatsu says, "our body is becoming more naive to bacterial pathogens."

Hiramatsu warns that with latent pathogens such as herpes turning aggressive by adopting genes, and new pathogens springing from the tropics, we can only be confident that our immune system will fight such enemies if it remains fully functional—if it matures properly when we are children and if it is not weakened later by misapplied drugs. Antibiotic and antiviral drugs have only been around for sixty years, a mere instant of human existence. It is too early to know whether the drugs that complement our immune system will have undermined the benefits of the natural selective pressure pathogens put on it. "Science may alleviate our sufferance from nature for a while," Hiramatsu says, "but the longer and better we protect ourselves in artificial ways, the greater the catastrophe may be once a novel pathogen arises, which could eventually cause extinction of our lives. This may sound like a stupid science-fiction story. However, in the future it would seem vital to evaluate how our artificial way of life influences our body's natural protection system."

So how can we maximize our ability to beat infectious diseases, without making them stronger and ourselves weaker?

At the absolute minimum, hospitals must clean up their acts. It's simply unacceptable that they have become among the most dangerous places on earth. To start, they must ensure that doctors and nurses clean their hands between patients and that infections are not transmitted from patient to patient by any kind of equipment—needles, bedside monitors, therapeutic machines, or surgical instruments. Administrators and regulators then have to clamp down on nursing homes and assisted-care facilities where nurses sometimes hand out drugs like candy and workers spread germs between elderly residents and the friends and family members who visit them. Otherwise, moving a senior citizen to these institutions could imperil them more than help them.

The other key way in which we can stop fueling microbial evolution is to end the overprescription of antibiotics. We can't stop organisms from developing resistance due to chance mutations, but we can de-escalate the arms race between us and them. As Stuart Levy at Tufts has been saying for years, doctors must prescribe antibiotics only when they suspect patients have bacterial infections, not viruses. And patients have to listen to doctors' advice instead of demanding nonexistent cure-alls. Doctors must also be careful about prescribing new drugs indiscriminately. For years, U.S. physicians used Cipro sparingly for only a few difficult infections, where it worked well. But after the anthrax scare they gave Cipro to thousands of people just to prevent them from the remote possibility of exposure, even though doxycyclene is just as effective. "There is still very little bacterial resistance to Cipro in the United States," Levy says, "but if we abuse it, resistance will grow, and we will have to cross it off our list of useful drugs." In addition, Levy says, governments in countries like Japan should stop allowing antibiotics to be sold over the counter, which encourages consumers to overuse them grossly.

Meanwhile, the FDA in the United States and regulators in other countries should stop farmers from routinely dumping antibiotics into animal feed, and consumers must be willing to pay a few cents more per pound of meat to cover the farmers' slightly higher costs. Consumers must also resist the temptation to buy magical-sounding antimicrobial hand soaps and other products that won't help them anyway; it is unlikely marketers will stop hustling them. These products just hurt the susceptible, nonthreatening bacteria, allowing the strong, threatening pathogens to run more freely. The public has got to be more wary of potions; by early April con artists on the Internet were advertising disinfectants that "kill SARS supervirus," dietary supplements that supposedly strengthened the immune system against the pathogen, and masks and gloves designed for the germ. These were patently ridiculous, since scientists had not even determined what the pathogen was yet, much less how to stop it.

Parents must appreciate the wisdom of allowing the natural powers of the immune system to develop fully in their children. While newborns and infants should indeed be protected from any and all germs, doctors say that once they have become toddlers they should not be artificially walled off from routine bacteria and viruses. Most immunologists would support letting children eat their toast that has fallen on the floor, as long as it hasn't landed buttered-side down. Their immune system is molded by its experiences; it must learn how to battle foes that will attack for years to come. Elementary school teachers will tell you they got sick frequently during their first year of teaching, but that in subsequent years they rarely were ill; their immune systems learned how to fend off the many organisms that children spread around.

The immune system can't develop properly if it isn't challenged when it is young.[2] Recent studies indicate that children who are less challenged by germs are more likely to develop subsequent allergies and asthma. Their T cells tend to make the kind of cytokines that help allergies and asthma develop. The same cytokines make it easier for TB and Epstein-Barr to thrive. For the rest of their lives these people will be at a disadvantage when it comes time to fight off serious diseases.

Philadelphia pediatrician Bonnie Fass-Offit adds that parents raising young children may have to change their mind-set and expectations. "Don't always look for a quick fix; there may not be one. Just know that when you have young children, especially if they are in day care with other kids, you will be spending a lot of time at home. Kids under age three get eight to ten viruses a year; plan on using all your sick days from work."

What we do want to guard against, of course, are the nastier bacteria and viruses that do severe harm. The single best defense we have is vaccines, which present safe challenges instead of deadly ones. Parents and schools must be rigorous about ensuring that all children get the full set of recommended vaccinations.

There is not much we can do about an unwitting carrier who happens to cough in our direction or greet us with a germ-ridden hand. But there is more we can do to reduce the pathogens that ride on our food. The number of cases of food-related illness each year is staggering. Safer food begins with food producers. We must demand that they cleanse the plants that process meat as well as fruits and vegetables and other products that spread *E. coli* O157, salmonella, listeria, and the rest of the long list of food-borne diseases. Congress must revise the mission of the USDA so that public health has priority over profits in the agricultural sector. Then legislators must give the FDA and the USDA the money and mandate to hire more inspectors.

Inspectors should have broader power to hand out swift and severe penalties or shut plants down, and they should be judged on whether they wield that power with impunity. Better quality control may mean consumers will have to pay a bit more for safer food, but we should let our legislators know we are willing to trade pennies for peace of mind.

Meanwhile, cooks and staff at all restaurants and fast-food outlets have to be better trained in how to handle food. Owners should be held accountable for assuring that their employees follow appropriate procedures. Otherwise more patrons like little Brianna Kriefall will die. Right now the chief levers seem to be lawsuits by victims, and bad publicity. But we all should get behind efforts of consumer protection groups like Safe Tables Our Priority, the Center for Science in the Public Interest, and the Consumer Federation of America to help balance powerful agricultural lobbies.

Given the sophisticated enemies our immune systems must fight, we are clearly better off preventing the pathogens from spreading in the first place. That means all nations must build stronger public-health systems. In underdeveloped countries, improving public health begins with cleaning up unsanitary living conditions: chemically treating public water supplies to kill basic germs and parasites, cleaning and refrigerating food supplies to reduce swarming bacteria, eliminating standing water to cut down on insects that spread numerous diseases, and installing blood-screening procedures to pull infected blood out of supply lines. The major human pathogens spread more easily in conditions of crowding and poor sanitation, which become worse as the population expands and poverty increases. Persuading people to practice safe sex and stop sharing drug needles will also help, and that's true for developed countries as well as developing and underdeveloped ones. Then national health administrators have to improve the public health infrastructure so that outbreaks of new diseases like SARS can be investigated and reported on a timely basis, and without political interference.

Vaccines must be made cheap so that poor nations can afford to inoculate their people. Although WHO was able to eradicate smallpox and is well on the way to eliminating polio, it has become clear that efforts to defeat AIDS, malaria, and TB exceed the available resources. Starting in earnest in 2000, a few public-private partnerships such as the Global Alliance for Vaccines and Immunizations have sprung up to bring together philanthropic foundations, pharmaceutical companies, governmental agencies, and the United Nations. The goal is to devise plans to improve health infrastructure, reduce the cost of drugs and vaccines, and develop needed new vaccines and pharmaceuticals. But much more is required from both public and private sources. U.S. voters, for their part, can let their public officials know how important it is that we keep our country's commitments to global health programs, and not withhold or cut funding to them.

If people living in developed countries think that they needn't bother trying to cure the third world, they should remember that the globe is much smaller than ever. Infectious diseases from other countries knock on our front doors every day. SARS was certainly a wake-up call in this respect.

Despite our best efforts, we will inevitably suffer outbreaks. But we can significantly limit them, as well as seriously cut down the spread of chronic diseases, with better public-health systems. The strong success of free clinics and DOT programs in curtailing TB not only in places such as Boston but also in Haiti and Peru indicates that health departments worldwide should apply directly observed treatment as much as possible against chronic and drug-resistant diseases. TB kills more than 2 million people each year, and most of that number contract drug-sensitive disease that could be effectively treated if social structures permitted. This requires governments that are willing to allocate the significant funds needed. Health departments, meanwhile, have to act wisely in tracking diseases, and remain vigilant in treating them. New York City blew it when it dropped its TB clinics in the late 1980s, thinking the disease was disappearing, and the national public-health system was slow to respond to the rise of HIV, deceiving itself that the disease was some sort of malady leveled by a vengeful God against the gay male community. The threat of bioterrorism and the recent experience with West Nile virus make it crystal-clear that we need to have a strong public-health infrastructure in place if we want to be able to respond effectively when natural or unnatural outbreaks occur.

To hasten communication about outbreaks—and thus improve our early-warning system—we've got to train physicians to better understand the symptoms they should be on the lookout for, and licensing boards should require more systematic, ongoing physician education about emerging diseases. Up-to-date knowledge does no good, however, if doctors are not more responsible about reporting patient cases. In New York City, doctors are supposed to report any of fifty-three diseases, but assistant health commissioner Marci Layton knows she hears about only a fraction of them. The new online networks developed by the CDC for tracking epidemics should make reporting easier, and help public-health officials get a more thorough and rapid handle on pathogens that are running around. Linking them all together in one global network could be the cornerstone of future public health, says Michael Miller, the CDC epidemiologist. "We learned with the anthrax attacks that better information technology could have filled a number of gaps" in gathering intelligence and issuing advisories to doctors on how to react.

The technology for improved reporting exists and it helped the CDC and the WHO respond quickly to SARS. What's needed is the will to make it happen. In 2002 the CDC awarded a $1.2 million grant to a Harvard University consortium to develop a model early-response system for bioterrorist attack. The computer system would automatically share information such as medical records from health departments and emergency rooms in all fifty states with the CDC, so that national authorities could detect unusual cases early, and electronically notify physicians around the country immediately. The grant is for a demonstration project; presumably the CDC will eventually be responsible for building and running a more comprehensive system.

The CDC has made some important changes, but there is much more to do. We should be aiming for the ideal scenario, where detecting and reporting of infections and possible outbreaks, and identification of new diseases or new strains, happens routinely in every corner of the country. Perhaps the Harvard system, combined with the CDC's expanding online networks and local health-department systems, will spawn such a web.

One key weapon needed for swift identification is better and faster lab tests, which would help doctors more quickly and accurately diagnose a patient's disease at its earliest stages. Jeannie Brown might have lived had she been diagnosed with group A streptococcus immediately. The problem is how to diagnose disease early, when most people show no definitive symptoms.

Fast action can separate life from death. When Seth Rogovoy, a music critic for the *Berkshire Eagle* newspaper in western Massachusetts, got out of bed on October 1, 2001, he felt pain in his right knee, which appeared red. By that evening his knee was tremendously swollen, it throbbed, and sickly red lines were radiating in an evil web down his leg. A fever suddenly soared. At 3:00 A.M. Rogovoy limped into Fairview Hospital sweating and dazed.

The doctors admitted him and gave him intravenous antibiotics, but by the third day Rogovoy's leg was purple from his hip to his foot. His heart rate was 140 beats per minute. Fairview transferred him to the county-wide Berkshire Medical Center. An infectious disease doctor put him on more antibiotics, drew some fluid from his knee, and sent it to the lab. For two days Rogovoy sweated out his unknown fate. Then the result came back: he had invasive group A strep. The doctors pumped antibiotics into Rogovoy for eight days while working down his fever, and finally beat the strep. It took three months of extensive physical therapy for his leg and knee to begin to recover.

Unlike Jeannie Brown, Rogovoy didn't wait until it was too late to seek help. And unlike Jeannie, he received fast infusions of antibiotics and timely lab test results. Many more rapid and inexpensive tests are needed to stop infectious diseases, and they must become inexpensive enough so doctors can routinely try many of them, instead of waiting for even two days. More research will result in more tests like those that can diagnose TB in a few hours rather than a few weeks or the new test for HIV that takes less than an hour. Rapid field tests that can sense bioweapons upon contact could also protect a city's population if the pathogens are released. Our need for more and better tests is underscored by the amazing statistic that 60 percent of organisms grown in a lab for testing cannot be positively identified.

Finally, and most fundamentally, the United States and other nations must step up basic research. We simply have got to know more about how pathogens work if we hope to craft better ways to stop them.

The first order of business is to make sure enough incentives exist so young people choose careers in both basic and applied research. Clinical research requires highly trained physicians, but fewer medical students are choosing that path. Blood expert and hepatitis C victim David Nathan recently chaired a high-powered panel at the National Institutes of Health that examined the state of clinical research, and he is worried.[3] "Clinical research is very hard work. You've got to be an excellent clinician and an excellent scientist. When I was coming up the line, we had only thirty or forty drugs, and nothing like the complicated tests today. Now, even in my own field, it's impossible to understand all that's happening. I wrote the major textbook on hematology—it's two volumes, and I can't tell you everything that's in them. Now a clinical researcher has to know molecular biology, molecular genetics, structural biology. Students get discouraged. How do we develop these people? If we don't have them, there will be no one to translate all the wonderful basic science into treatments that benefit real people."

Training to go into clinical research adds years onto the already long period it takes to become a physician. These young doctors then usually must endure low-paying fellowships for several years, just when they are struggling to pay off medical school debts. They may spend another few years working semi-independently in someone else's lab. Finally they are ready to try to find an independent job either in a medical school or at a clinic; at the moment these jobs are in short supply.

Nathan suggests that the government institute a debt-relief program for young physicians who pursue research, to "relieve anxiety and improve morale in the ranks of nascent clinical investigators." He further suggests that grant-givers who make awards to established researchers earmark more money within the grant to go to salaries for young doctors who will be doing part of the work while they are transitioning to an independent career.

Veteran researchers have their own hurdles, too. At almost all colleges, faculty are expected to generate their own funds for research. This means coming up with a novel research proposal, and shopping it against other proposals at a governmental agency such as the National Institutes of Health or private foundations for funding. Most grants last three to five years, then the researcher must conduct a new search for funds to keep the work going. In the best years the NIH funds only 30 to 40 percent of the competing applications they receive. In tight-budget years the awards drop below 20 percent. To better sustain work, Nathan suggests institutions award funding for equipment that can be pooled among different researchers or colleges, and grants that support group efforts, to lessen the burden on any one person.

Money is not the only challenge. Researchers must now also worry about losing institutional support. Nathan cites a case at the

University of Toronto in 1996. A doctor, Nancy Olivieri, was experimenting with a new drug that promised great improvement over older drugs to treat thalassemia, a rare blood disorder. But the drug turned out to be troublesome—it caused hepatitis. Olivieri wrote a paper to warn others, and the company that made the drug promptly sued her for defamation. The administrators at the Hospital for Sick Children, where her experiments took place, not only did nothing to back her up, they claimed she had not behaved professionally. (The hospital's allegations were later rejected in their entirety by an independent board.) Her school provided no help. It turned out that the company was supporting the school, and had offered millions of dollars to the hospital. It took until November 2002 to settle the dispute, in Olivieri's favor. "We can't do clinical research without industry," Nathan says, "but everybody has to behave properly. The patient has to come first, second, third, fourth, and fifth."

Bringing more antibiotics and antiviral drugs to market is a thornier problem. Pharmaceutical companies must conduct more research to refill the pipeline of new drugs, which has stalled to a trickle. This requires overcoming the industry's troublesome model of looking only for blockbuster drugs. Streamlining the regulatory approval process, capping lawsuits by patients, and protecting companies unless they are negligent could help make drug development more feasible for small companies as well as large ones.

Universities and drug companies are the primary places where biological science progresses, and the more work we can do, promptly, the better. Our war with infectious diseases is constantly escalating. They attack. We defend. They counterattack. We counterdefend. But they have two fundamental advantages: They vastly outnumber us, and they mutate faster than we do. We are always a step behind. Hepatitis C and HIV have mastered the ability to outrun our immune system, in part by reshaping themselves faster than we can reshape our immune cells.

Our best hope of leapfrogging their advances is to pin down the basic workings of each pathogen. The research can be incredibly arduous and time-consuming. It took Rod Welch and Fred Blattner six years at the University of Wisconsin to sequence the *E. coli*

O157 genome. Hiramatsu's large team in Japan needed an entire year to sequence methicillin-resistant staph. And drafting a blueprint of a pathogen's genome is only the first step. Just moving from understanding *E. coli*'s genome to constructing a full model of how the bacteria functions is expected to take hundreds of scientists ten years and cost at least $100 million.[4] The money will be well spent, because the work will establish a means for deciphering how a host of other pathogens work, too. But we could certainly use the results sooner. *E. coli* O157 alone will sicken and kill many people in the interim.

One exciting offshoot of understanding better how organisms work is the ability to harness their tricks to our advantage. Getting drugs past the gut, into the bloodstream, and then into the right target cells is always a problem. A compound may work well against an organism trapped in a lab dish, but many times there's no way to deliver that drug effectively to the enemy running free in the human body. Scientists working with HIV noticed that one of its important regulatory proteins, called tat, was masterful at sneaking into human cells. Exploiting that observation, the scientists found that if they attached a drug or protein to tat, it would drag the substance along with it into the cells. Researchers at Stanford University are now trying to learn how tat does this, so they can exploit the mechanism to improve drug delivery.

Other researchers are trying a variant of cholera's toxin to usher antigens efficiently from an orally administered vaccine through the gut so that immune cells will recognize them.[5] Still others are experimenting with canary pox and modified HIV viruses to deliver vaccine antigens, as well as to get cancer-fighting genes into tumor cells and anti-inflammatory genes into arthritic joints. Ironically, David Knipe at Harvard Medical School is investigating how herpes might be engineered so it can still infect cells but can't reproduce.[6] This would allow him to use a modified herpes 1 virus—the cold-sore type—to act as a vaccine against the closely related herpes 2, the genital lesion type. This would be a case where a patient allows a doctor to inoculate him with a less dangerous virus to trigger immunity against a more dangerous one. If it worked, we would be tricking our enemies for a change.

Bacteria themselves could be a source of new antibiotics. Bacteria protect their turf by releasing small proteins that kill other bacteria. Researchers in the Netherlands are grappling with a molecule called nisin made by certain strains of *Lactococcus lactis*, the bacteria that turns milk sour but does not harm us. The nisin compound can kill a wide range of other bacteria.

There is so much promise to be found with more research. Conversely, it is shocking to realize how little we know about our microscopic enemies. For years doctors told us we should lower cholesterol and exercise to prevent heart disease. Then, in the late 1990s, a study of risk factors for heart disease unexpectedly showed that people taking tetracycline and quinolone antibiotics had fewer heart attacks. Why? Could infections be contributing to heart disease? As scientists explored that possibility, more circumstantial evidence accumulated—people who had herpes 1 or chlamydia pneumonia were more likely to develop heart disease. Other researchers found evidence that macrophages played a major role in inducing the artery-clogging formation of plaque by cholesterol.[7] Most recently the level of C-reactive protein, a protein made by the liver in response to inflammation caused by infection, was found to be an even better predictor of heart disease than cholesterol. Researchers have also now determined that infections are at least partly responsible for certain forms of diabetes, multiple sclerosis, and arthritis. We also have learned that the papilloma virus causes most forms of cervical cancer.[8] If we have missed such basic traits of infectious pathogens, we can have little confidence that we know how to defeat them.

More research, a more dynamic public-health system, and many of the other steps we must take to improve our odds require greater funding, much of it from government sources. That, of course, puts the fight against infectious diseases in heavy competition with numerous other social and health-care causes. But we must factor into the equation the longer-term savings to be gained by more concerted action. We would be saving countless lives.

Local health departments—the institutions most likely to protect each of us in our hometowns from outbreaks—simply need more money from the cities and states that oversee them, and from the CDC and other federal agencies that contribute to specific programs. In 2002 the Institute of Medicine issued a report concluding that the public health system was badly in need of improvement. It found that 30 percent of health departments don't even have e-mail. It also found a lack of credentialed experts. Institute officials said not only the government, but also businesses, the media, and community organizations needed to get involved to improve the situation.

The health departments know what to do; they just can't pay for it all. New York City's assistant health commissioner Marci Layton says, "Our infrastructure has been very fragile for a very long time. It's all determined by funding. In New York City, AIDS, TB, and sexually transmitted diseases have their own programs. But the rest of what we fight is lumped together under one wastebasket term—'emerging infectious diseases' or 'communicable disease'—and funding for it has not been very forthcoming. We really should be dealing with these threats disease by disease."

Money for research comes almost entirely from the federal government, and mainly from the NIH. In 1998 Congress agreed to double the NIH's budget over the subsequent five years. Over the previous four decades, Congress had doubled the budget roughly every ten years, and the increases were anything but smooth. During the late 1980s and early 1990s, the annual NIH appropriation became unpredictable, partly because of the federal deficit and a recession. The number of new grants awarded dropped precipitously and at random, and many ongoing grants that had been awarded earlier were cut. The uncertainty had a destabilizing and demoralizing effect on the research community. The five-year doubling of the NIH budget that began in 1998 has been a welcome boost. But the scheduled increases end in 2003. The question now is, what will Congress do next?

The outlook early in 2003 was far from certain. Rather than sustain the 14-to-16-percent annual increases that NIH enjoyed during the five-year expansion, President Bush's budget requests at that time for fiscal years 2004 to 2007 sought increases of just 2.1

to 2.3 percent annually. That would merely cover inflation—if it stayed low. Research inevitably becomes more expensive over time as new technologies take hold, more diseases emerge, and outbreaks become more serious and widespread. A recent report estimates that it takes at least a 6-percent increase each year to keep NIH from losing ground.[9] Lower increases also reduce its flexibility; because grants are typically awarded for three to five years, a large part of the budgets for future years is precommitted. In past years, Congress could be counted on to increase appropriations beyond low presidential requests, but there is fear that with the current budget crisis, NIH and the research community could be in for trouble. We also must be careful that spending more to counter bioterrorism doesn't actually divert money from research on the multitude of infectious diseases we know we will face. Paying the huge tab could even siphon money from other NIH research programs to counter fundamental illnesses such as heart disease and lung disease.

The answer is not to fight over how to divide up the existing research pie. We must create a bigger pie altogether. That is the only way we can defend ourselves against bioterrorism as well as the rise and reemergence of evolving infectious diseases. The future costs of not doing so—both human and economic—will be so much higher than the money we would have to earmark now. The case of AIDS proves that point all too well; if we had made AIDS prevention a priority and made drugs available at lower costs earlier, the epidemic would not have gotten so horribly out of control around the globe.

Surely this country is rich enough that we can find the funds to bolster our defenses against the disease threat. Fortunately there are congressmen who understand the importance of basic medical research, with Senators Tom Harkin of Iowa and Arlen Spector of Pennsylvania playing leading roles. Representative John Porter of Illinois and Senator Connie Mack of Florida were instrumental in creating the climate in which doubling the NIH budget became a priority.

Greater public support for such measures, as well as louder public objections to cuts in research, public-health budgets, or safety regulations, are vital. Public apathy, even if it is just perceived apathy, will quickly kill initiatives that aim to do more. We cannot allow ourselves to be lulled into a false sense of security. Most of us think of infectious diseases as plagues of the past or a problem of developing countries. Their sinister ghosts spring to mind when we see rows of gravestones bearing the names of children who died from polio or whooping cough, or hillsides of markers for adults killed by rheumatic fever or flu. In those bygone decades, even a healthy person could expect to die at any time. Today in the United States we expect we will be able to defeat most infections, at least while we are young and healthy.

Yet confidence in the medical and research communities that our immune systems, the pharmaceutical industry, and modern medicine are adequate to combat the pathogen threat is slipping. The sense of challenge was palpable among a small, hushed crowd of microbiologists who sat in a dimly lit conference room for a quietly organized meeting at Harvard Medical School in September 2000. A handful of the nation's leading microbiologists presented their latest findings. One by one they explained what they knew. And one by one they impressed the group with what we still don't know. The last presenter was Hidde Ploegh. He confidently explained new computer-generated diagrams showing some novel escape mechanisms that herpes employs. But when he finished, he turned squarely to the audience, humbly raised his arms, and said with uncharacteristic chagrin, "Really, we haven't the foggiest notion how some pathogens work."

Our fate will be decided by how fast we act. At the same symposium, C. J. Peters, then chief of the CDC Special Pathogens Branch, told the insiders that his agency was bracing for a rise in domestic outbreaks of infectious diseases. The invaders were not waiting for science to figure them out, he said. We are in a race against time.

EPILOGUE

THE NEW NORMAL

The global village is a small and infectious place. Consider what has happened in the year that elapsed between the hardcover publication of *The New Killer Diseases* and this paperback edition:

• Monkey pox appeared in the United States for the first time, infecting dozens of people in the Midwest. Symptoms include skin lesions, muscle pain, fever, and possible brain swelling. The virus was traced to pet prairie dogs that had been infected by an exotic imported pet known as the Gambian giant rat.

• Mad cow disease, at one time confined to the United Kingdom and Europe, struck a cow in Canada. Scores of countries, led by Canada's suspicious neighbor, the United States, banned the import of Canadian beef, crippling that country's industry. Then a mad cow turned up in an American herd, and the world placed a moratorium on U.S. beef that would cost $3 billion a year if it had been sustained.

• Several people died from a West Nile–like sickness called eastern equine encephalitis, which breeds in horses and kills half the humans it infects. The virus killed a man in Georgia, infected horses from Florida to the Carolinas, was found in mosquitoes as far north as New York, and sickened a two-year-old girl in New Jersey—the first time a human case had occurred in that state in twenty years.

• The number of HIV cases in the United States rose at a rapid 5.1 percent annual rate, largely because people at risk became less cautious, perhaps due to the misconception that HIV can easily be contained with the right combination of pills.

• In Chicago, contaminated green onions imported from Mexico caused a hepatitis A outbreak that sickened more than 500 people. Earlier outbreaks also traced to Mexican green onions had been reported in Georgia, North Carolina, and Tennessee.

• The flu season came early and hard.[1] By December 5, 2003, 6,300 cases had emerged nationwide, more than the number that had appeared by the same date in the prior two seasons combined. People rushed to get immunized, but found vaccine to be in short supply.

Not that the vaccine would have necessarily helped. Health officials told the public that the flu shots they were clamoring for did not even include the Fujian strain, which ended up being responsible for three-quarters of the 2003 influenza cases; the Fujian flu had emerged too late to be added to the vaccine. The U.S. Centers for Disease Control, the World Health Organization, and other authorities had done their best, as they do every year, to predict which strains would threaten the most people in the Northern Hemisphere, but in 2003 the best combination eluded them.

If all that wasn't enough, in January 2004 the WHO warned of an "unprecedented" simultaneous appearance of avian influenza—bird flu—in chickens, geese, and ducks in ten Asian countries. By late March, Vietnam and Thailand had reported thirty-four human cases, twenty-three of which ended in death.[2] The strain, A(H5N1), was of the same general type that killed a Hong Kong boy in 1997. Hong Kong slaughtered 1.5 million chickens and geese after that boy's death, only to see a similar pathogen raise its ugly head six years later. The 2004 outbreak killed nearly 70 percent of the people it infected. Thailand's leaders called out the army to help in the mass slaughter of fowl.

In February two Delaware farms reported outbreaks of a different and generally less dangerous form of bird flu. Japan, South Korea, Malaysia, and Singapore quickly halted the import of U.S.

poultry. A week later the same U.S. avian flu turned up at several markets selling live chickens in New Jersey. Just as that outbreak was brought under control, health officials detected a third bird flu type in chickens in Texas. In March an even larger outbreak was found in Maryland. Two independent reports were published showing that the "Spanish flu" virus that killed 500,000 Americans in 1918 was similar to the avian flu virus that had infected dozens of people in Asia in early 2004.[3] Was it only the fast response and slaughter of millions of chickens that saved us from another catastrophic pandemic? No one knows.

Steps have been taken over the last year to contain new outbreaks, but there is much more that needs to be done. Consider the United States' handling of its first case of mad cow disease.

Safeguarding foods has always been a great challenge, as has been outlined in earlier chapters. In May 2003 Canada reported that a cow in northern Alberta had been stricken with bovine spongiform encephalopathy, or BSE. That country's beef industry was soon crippled as more than thirty of its trading partners banned the import of Canadian beef.

The first U.S. cow with BSE was detected on December 23, 2003, in Mabton, Washington. The six-year-old had been imported from Canada in August 2001 along with eighty other cows. The fact that the fateful cow came from the same region as the Canadian BSE case makes it likely that contaminated feed made from the ground-up remains of previously slaughtered infected cows was the source of the infectious prions in both cases—the same mechanism that spread mad cow disease across Britain.

As of February 2004, the USDA had only tracked down twenty-nine of the eighty cows in the Canadian shipment containing the known infected cow. Most of them were found in Washington, and one in Idaho, but fifty potentially infected cows remained at large. The trail soon went cold and the investigation was closed.[4]

Despite the scare, the USDA made no move to mandate universal BSE testing for cattle or even to test all downers (cows that can't walk due to disease or injury). As early as 2001, the European Union had ordered mandatory testing of all slaughtered cows older than thirty months (BSE is almost always detected in older cows, so it was thought that relatively young cattle would be safe), and it picks up hundreds of infected cows each year, particularly in the United Kingdom, France, and Spain. Japan also began testing every cow slaughtered for human consumption in 2001, and it has since found prions even in the brains of several animals younger than thirty months. As of March 2004, the USDA was still testing only 40,000 out of 35 million animals killed annually and had announced a "big" increase to 201,000—still a small fraction of the total that go to supermarkets.

A disturbing report from Turin, Italy, in March 2004 suggested that the connection between prion infection of cows and human disease might be older and broader than previously thought. Researchers found evidence for a second type of prion that infects cows and causes a milder disease than mad cow. The new prion had a number of similarities to one of the variants that causes a naturally occurring prion disease known as sporadic CJD in people.[5] It had previously been thought that the disease arose spontaneously, afflicting about 300 Americans a year, but the Turin report raises the possibility that at least a subset of those cases might be linked to the consumption of infected beef. To make matters worse, the first case of probable human-to-human transmission via blood of "variant" CJD—the human form of mad cow disease—was reported in England in early 2004.[6]

A recent eye-opening USDA study found that 35 percent of ground beef samples included spinal cord tissue—after the brain, the part of the cow most likely to transmit BSE to humans if ingested. Unfortunately, the U.S. infected cow had already been processed by the time the test results were available. Thirty-eight thousand pounds of beef were recalled, but by then a USDA official estimated that Americans had eaten as much as 17,000 pounds of potentially contaminated meat.[7] Stanley Prusiner, the University of California, San Francisco, microbiologist who discovered prions, said recently that the USDA has been willfully blind to the threat of BSE and should have been testing far more animals ever since the British problem became clear in the late 1980s.

In light of the mad cow scare, U.S. Secretary of Agriculture Ann Veneman appointed an international panel of experts to recommend measures against BSE. The panel reported in February 2004 that the United States was not up to internationally recognized standards. Panel members were flatly unimpressed by the tenor of the federal government's response to the U.S. scare—that since the animal had been imported, nothing suggested any of the U.S. stock was unsafe. The panel recommended that no animal protein of any kind be fed to cows and that animal identification and tracking be significantly improved.

In January 2004 the USDA imposed new rules to ban the feeding of cow's blood to calves and chicken waste to cows; the blood could potentially contain prions that would infect the calves. Because there is no restriction on including cow remains in chicken feed, it would theoretically be possible for chickens to eat contaminated feed and eliminate prions in their waste, which could then infect cows that ate the chicken waste. We don't know for sure that this can happen, and there is no evidence that chickens themselves can become infected, but it might be better to be safe rather than sorry. The new rules also say that material from downers can no longer be used for cosmetics and dietary supplements. The brain, skull, eyes, and spinal cord of animals thirty months or older, and the tonsils and parts of the small intestines of all cattle, will be banned from human food. Also now forbidden is mechanically separated meat, which has been recovered from bones after slaughter using a scraping machine that can inadvertently break off chips of the spinal column that include nerve tissue. These steps close some but not all sources of risk to people.

The government's reluctance to admit danger for the sake of protecting an industry is one of several lessons that once again remained unlearned in the past year, even beyond mad cow. For example, the debate over dumping low-dose antibiotics into feed for animals such as cows, pigs, and chickens still rages on, even though we now know that the antibiotics strengthen drug-resistant bacteria both in the animals and in the people who then eat them as meat. WHO officials have advised that the practice be discontinued, after further research under a far-reaching program in

Denmark; that country banned the use of such drugs in 1998. Recent follow-up showed that the cost to pig farmers related to lost livestock or lesser growth increased by only 1 percent, yet the number of antibiotic-resistant bacteria in people's guts dropped dramatically. The European Union has decided that member countries should stop adding antibiotics to feed by 2006, yet the FDA is still dragging its feet; it has not forbidden any drug additives and has said it will only consider banning new drugs on a case by case basis.

The cattle industry has successfully fought tough regulations for years and seems to still hold sway over regulators. The USDA backed a two-year delay in implementing a plan to label meat with its country of origin, which Congress passed in 2002. The delay came in response to complaints from meat processors and supermarkets. And even if new regulations were put in place, would they be adhered to? Other feeding bans have not been rigorously enforced in the past, and some companies have found that violating bans, even repeatedly, has not led to penalties. The only way to ensure implementation and enforcement of any new regulations is for consumers to loudly voice and demonstrate their concern.

Given the number of infections that start in animals and cross the species barrier into humans, global surveillance of animals clearly needs to be improved. Thankfully, animal pathogens often do not have the cellular machinery necessary to spread readily among people; however, in the rare instances that they do, humans are likely to have little defense against them. Experts widely believe that species jumping led to the 1918 influenza pandemic, as well as the 1957 "Asian flu" that killed 70,000 and the 1968 "Hong Kong flu" that killed 34,000. Mad cow disease does not appear to have infected any humans in the United States yet, but between 2003 and 2004 there were several instances of other infections that jumped from animals to humans.

The Midwest monkey pox outbreak was traced to a shed in a Chicago suburb from which a man conducted wholesale trade in prairie dogs, Gambian rats, opossums, degus, and mice, apparently legally.[8] The Department of Agriculture has no restrictions on the importing of numerous animals, from reptiles, monkeys, and lions to chinchillas, mongooses, and hedgehogs. Most states and local laws allow pets to be traded freely or don't address the subject at all. This lack of oversight could have dire consequences, not just for humans, but for our pets and native wildlife. The exotic pet business is worth billions of dollars worldwide. Dealers in Texas alone sell more than 20,000 prairie dogs annually. If sick prairie dogs are released into the countryside at any point, the critters could start an epidemic among native animals, spread it to squirrels, and perhaps threaten millions of domestic dogs and cats.

Legal animal markets pose enough of a danger. Illegal animal trade raises the stakes. The U.S. Fish and Wildlife Service has a hard time keeping up with illegal wild animal meat, including flesh from rats, bats, and turtles as well as monkeys. Last year Boston officials seized the remains of twenty-six butchered monkeys that a passenger from Guinea had flown in for a wedding reception (such fare is considered a delicacy in some cultures).[9] The number of seizures at Boston's Logan Airport has risen substantially in the past three years.

More and more data suggests that SARS originated in animals, too. The virus most likely jumped to people from exotic animals that are part of the Chinese cuisine, with civet cats being the leading suspect. Tests in China showed that 40 percent of wild animal traders and 20 percent of wild animal butchers had developed antibodies to SARS—meaning they had been infected, even though none of them reported being ill. The virus was found in high concentrations in dried civet cat feces and might have infected people as they walked through live animal markets, kicking up and inhaling cat feces that had fallen to the ground.[10]

Animals may also have played a role in the large outbreak at the Amoy Gardens apartment building in Hong Kong, where more than 300 people developed SARS in a single month. Investigators found that several cats, a dog, cockroaches, and rats there were either infected or carried the SARS virus.

Labs around the world, coordinated by the WHO, figured out what SARS was in record time because of a crash program in March 2003. Yet even with that effort, a key finding did not come to light until January 2004. The virus had begun its ascent in China in November 2002 but smoldered until February 2003, when it ran rampant across China and then the world. In a paper published online in *Sciencexpress,* the Chinese SARS Molecular Epidemiology Consortium finally figured out why. Its study of virus samples showed that mutations in a single protein on the virus's surface—a so-called spike protein that the virus uses to attach to target cells—is what changed SARS from a pathogen that did not spread well from person to person to one that was highly contagious. By February 2003 the spike protein had mutated to its more effective form.[11]

A change in a surface protein is the very mechanism scientists believe was behind the great pandemics of the past, and it is what they now worry could allow the Asian bird flu to recombine with a strain of human flu to make it contagious among people. Flu expert Robert Webster, who heads the WHO collaborating center at St. Jude's Children's Hospital in Memphis, has warned that the Fujian flu mixing with the avian flu virus could create a virus with the worst properties of both.

Simple regulation and enforcement is the answer for preventing the spread of disease from animals to humans. Veterinary tracking also must be improved and integrated into public health networks, and medical and veterinary schools should work much more closely with one another. Medical schools have better resources for studying infectious diseases and could lend a hand in studying animal pathogens. And if medical researchers were more aware of animal pathogens and tested for them in humans, perhaps they could identify new human diseases sooner or more accurately.

When the avian flu first sprung up in 2004, WHO spokesman Dick Thompson told the media that there are "big holes in the global public health network" for monitoring the many animal diseases that have implications for humans. Public health leaders are even proposing a global monitoring system for emerging animal diseases that mirrors the WHO network for charting the continuing rise of novel human influenzas. Unfortunately, global health budgets are already stretched so thin that it seems unlikely such a proposal will be funded, except perhaps for specific diseases like bird flu that have immediate implications for both agribusiness and human health.

The widespread SARS epidemic, as well as the sudden appearance of mad cow, make it plain that health agencies (and Congress if needed) should mandate local, state, and national reporting of many diseases that now are only tracked piecemeal, if at all. When SARS emerged globally in March 2003 China was harshly criticized for not advising the rest of the world that a new threat had appeared, and top officials were fired. And yet, when Chinese media began to report on what sounded like a new bird flu in early 2004, the government tried to silence those news outlets. The current practice of reporting disease all over the globe, including within the United States, is haphazard and provides no systematic way with which to detect emerging threats.

If a global health monitoring system seems far away, there still are measures we can take to protect ourselves, such as developing new vaccines. Three labs in the WHO network are working on a vaccine for Asian bird flu; researchers are essentially reverse-engineering the virus's genes to figure out how it functions and how to design a vaccine that could stop it. Three other groups, one in Israel and two in the United States, are investigating mice to see if giving them antibodies to West Nile virus after they have been infected helps reduce the severity of illness. So far the antibodies seem to help prevent death if given early enough. NIH's Laboratory of Infectious Diseases also showed that a new vaccine made from a crippled dengue virus protected rhesus monkeys from West Nile.

The National Institute of Allergy and Infectious Diseases gave a $36 million grant to Chimerix in San Diego to develop an oral drug that could possibly treat smallpox if a victim takes it within three days after being exposed. It would be intended for people who were not vaccinated or who have compromised immune systems and therefore have been advised to not take the standard smallpox vaccine because it could lead to serious complications.

The institute has also begun early human trials of a vaccine against the dreaded Ebola virus, which has killed people in gruesome fashion during outbreaks in Africa and is also decimating the great apes there.[12] The experimental vaccine, made by Vical in San Diego, uses synthesized DNA based on modified Ebola genes, and therefore does not contain any infectious material from the virus itself. The U.S. Army Medical Research Institute for Infectious Diseases is designing a second vaccine based on particles of viral proteins that also shows promise.

However, vaccine development is only one piece of the puzzle. Educating the public about the effectiveness and side-effects of vaccines—which dangers are real and which are imagined—is also important. Concerned parents continue to stubbornly spurn longstanding childhood vaccinations for their kids—reopening the door for ugly scourges despite even more studies proving that vaccines are safe. Opposition to vaccination by Islamic leaders in Nigeria, on the baseless fear that it could cause infertility in girls, allowed polio to spread to four neighboring countries that had eliminated the disease. A measles epidemic cut through Campania, Italy, due to inadequate vaccination coverage. American mothers and fathers were just as foolish: For example, health officials in Westchester County, New York, blamed parents who refused to have their children vaccinated for causing an outbreak of whooping cough that struck seventeen children and two adults.[13]

Measures must be taken to improve the vaccine supply chain as well. Drug companies are pulling out of vaccine manufacturing because profits are low, and it is becoming more and more evident that the exodus is leaving little margin for glitches in production. In the past three years, U.S. shortages occurred in eight of the eleven standard childhood vaccines, and in late 2003 the National Vaccine Advisory Committee told the Department of Health and Human Services that those disruptions are likely to continue. The committee recommended that the government expand stockpiles, strengthen liability protection for manufacturers against frivolous lawsuits, and create financial incentives for manufacturers. The National Academy of Sciences advocated that the government underwrite vaccine development and provide subsidies for people who are uninsured and can't afford inoculations.

For diseases we do not yet have vaccines for, we can best protect ourselves by continuing investigative research and developing better diagnostic tools. An accurate blood test for West Nile virus was instituted in July 2003, for example, and it proved to be an important step.[14] An estimated twenty-three people were infected through blood transfusions during the summer 2002 virus season. In 2003, the new test detected 601 contaminated donations in the United States. Only two people became infected with West Nile from a blood transfusion during the 2003 season.

Scientists are devising similarly rapid tests for SARS, so patients suspected of being infected don't have to wait days for reports to arrive from outside labs. The current "PCR" test that can detect SARS's genetic material early after infection only picks up 54 percent of cases. The antibody test now used identifies 96 percent of cases but only when it is used after a patient's immune system has begun responding to the infection, which can take as much as three weeks—very late in the game to stop the person from spreading the virus to others. We need a test that can spot SARS in its early stages with high reliability. Faster results will help doctors and public health officials better contain the spread of SARS by people who are infected and don't know it or who are awaiting diagnosis.

As the success of vaccines and diagnostic tests show, prevention and early detection are the best ways to keep ourselves safe. That concept applies as urgently to protecting ourselves from bioterrorism too. And yet, the Council on Foreign Relations noted that the United States is not spending enough to prepare first responders such as emergency medical teams, firefighters, and police to handle bioterror attacks. It also found that 75 percent of state disease labs are already overwhelmed under normal conditions, and therefore would not have the capacity to act quickly in a crisis.

One event that lent credence to the idea that we are still not prepared to meet a bioterrorist threat was a classified, war-games-style exercise dubbed Scarlet Cloud, overseen by the Homeland Security Department, the Transportation Department, and the White House Homeland Security Council. In this simulation, one similar to the Dark Winter exercise for smallpox in 2001, anthrax carried in different types of aerosols was released in several American cities. Government agencies in virtual responses were unable to distribute and administer antibiotics quickly enough to prevent large numbers of deaths. Although the mock drill showed that the nation is better prepared to respond to an attack than it was before September 11, 2001, the Homeland Security department said work must still be done to make the antibiotic supply chain more efficient.

To improve the situation, the Federal Bureau of Investigation is analyzing whether to propose a dedicated, nationwide microbial forensics system that would have the expertise and infrastructure to hunt down the source of biological attacks, perhaps through tight networking of existing federal, state, academic, and private-sector labs. NIAID awarded two grants of more than $100 million each to Boston University Medical Center and the University of Galveston to build "biosafety level-4" labs—the most secure and safe possible—for handling the most dangerous pathogens. NIAID awarded nine other institutions lesser amounts to build biosafety level-3 labs, which will be called Regional Biocontainment Laboratories. And the Department of Homeland Security started Project Bioshield to establish an infrastructure for routinely testing environmental samples for bioterrorist threats. Another step in the right direction could come from President Bush's February 2004 call to significantly increase the number of doctors and nurses in public health services; the goal is to help economically depressed areas in normal times and swell the ranks of medical personnel who could respond in times of emergency.

Recent outbreaks put in sharp relief some significant actions that doctors, researchers, public health officials, and politicians should take. Chief among them is ensuring greater surveillance and study of infectious diseases. As Congress and other authorities with budget power address such measures, they must continue to allocate appropriate funds for research. This is where our greatest hopes lie for prevention and cures. Researchers are making progress; we need them to continue at an even greater rate. SARS, Lyme disease, *E. coli,* tuberculosis, hepatitis C, herpes, and a host of other new and reemerging infectious diseases continue to challenge us. The human race has been under a sustained siege for more than a decade now, and the trend shows no sign of slowing. As Julie Gerberding, director of the CDC, has been saying repeatedly during public appearances, the steady microbial attack is "the new normal."

While Gerberding's words reflect a bit of melancholy, they also serve to counter hyperactive media coverage. A year ago newspapers and television networks were being accused of fueling hysteria over SARS and other ills. Yet the headlines have continued so unabated about one new threat after another that health experts now worry that people may be turning a deaf ear. That could undermine what had been healthy, growing pressure on government to improve public health measures. Gerberding says the country has got to focus, expand research, and fund far more work to unravel how pathogens function and how to defeat them—a call to arms that has been made throughout this book. We have to raise the bar if we are to meet the new normal head on.

Tighter surveillance, more research, better tests, expanded public health programs, tougher policies, and greater education will all help us stay a step ahead of infectious diseases. But these measures require more money and change, neither of which is likely to be embraced unless elected officials and regulators feel more pressure from the public. When President Bush proposed his 2005 budget, it included a big boost of 15 percent for Homeland Security applied science, but only a 2.7 percent increase for the NIH. And most of that meager increase was committed to areas such as biodefense. If steady battle with infectious diseases is the new normal, then we must press politicians to boost science and public health to a higher level—to its own new normal, where researchers, administrators, and elected officials are doing as much as they can. So get informed, get agitated, and get active.

April 2004

NOTES ON SOURCES AND FURTHER READING

This book is based on interviews with many people on the front lines in the battle against infectious disease. Scientific journals and news sources also were used extensively, along with key websites, particularly that of the CDC. The notes that follow contain selective references readers can go to for additional information. For a more comprehensive background, consider textbooks such as P. Murray et al., *Medical Microbiology,* 4th ed. (St. Louis: Mosby, 2002), for basic medical microbiology and infectious diseases, and I. Roitt and P. Delves, *Essential Immunology,* 10th ed. (London: Blackwell, 2001), or C. Janeway et al., *Immunobiology: The Immune System in Health and Disease,* 5th ed. (New York: Garland, 2001), for immunology.

INTRODUCTION

1. M. Enserink, "A second suspect in the global mystery outbreak." *Science* 299 (March 28, 2003): 1963.
2. www.CDC.gov for webcasts and news conferences, virus structure, and ongoing data about cases and fatalities.
3. "Disease's pioneer is mourned as a victim," Donald G. McNeil Jr., *New York Times,* April 8, 2003.
4. CDC SARS Investigative Team and M. Charles. "Severe acute respiratory syndrome (SARS) coronavirus testing—United States, 2003." *MMWR* 52 (April 11, 2003).
5. "Lab decodes genes of virus suspected in lethal outbreak," Donald G. McNeil Jr., *New York Times,* April 14, 2003.

6. M. Hamburg, J. Lederberg, B. Beatty, et al. "Microbial threats to health: Emergence, detection, and response," Institute of Medicine, National Academy of Sciences. National Academies Press, March 2003.

7. M. Enserink, "WHO wants 21st century reporting regs." *Science* 300 (May 2, 2003): 717.

8. www.who.int for case numbers and history of SARS outbreak.

CHAPTER 1

1. Author interviews with Audrey Trantham, Sherry Trantham, Vivian Howard, Michael Keogh, and Kathryn Freyfogle.

2. Website for the New York City Department of Health Bureau of Communicable Diseases, www.ci.nyc.ny.us, and group A streptococcal disease/technical information at www.cdc.gov.

3. "Summary of notifiable diseases, United States, 1998," *Morbidity and Mortality Weekly Reports* 47 (Dec. 31, 1999): 78–83.

4. National Intelligence Council report, "The global infectious disease threat and its implications for the United States," Jan. 2000, at www.cia.gov.

5. See www.cdc.gov. for information on Lyme disease and malaria.

6. See www.cdc.gov for information on West Nile virus.

7. Agents/HFV at Center for Civilian Biodefense Strategies at www.hopkinsbiodefense.org.

8. See www.cdc.gov for information on food-borne pathogens.

9. "Outbreak of invasive group A streptococcus associated with varicella in a childcare center—Boston, Massachusetts," *Morbidity and Mortality Weekly Reports* 46 (Oct. 10, 1997): 944–48.

10. E. Levy and T. Monte, *The Ten Best Tools to Boost Your Immune System* (New York: Houghton Mifflin, 1997).

11. "The Global Infectious Disease Threats": National Intelligence Council.

12. C. Harvell, C. Mitschell, J. Ward, et al., "Climate warming and disease risks for terrestrial and marine biota," *Science* 296 (June 21, 2002): 2158–62.

13. "Pertussis—United States, 1997–2000," *Morbidity and Mortality Weekly Reports* 51 (Feb. 1, 2002): 73–76.

14. See CDC website www.cdc.gov. for information on influenza.

15. J. Fauber and M. Johnson, "Did wild game feasts lead to fatal brain disorders?" *Milwaukee Journal Sentinel,* July 21, 2002.

CHAPTER 2

1. J. Cello, A. Paul, and E. Wimmer, "Chemical synthesis of poliovirus," *Science* 297 (Aug. 9, 2002): 1016–18.
2. Public Health Emergency Preparedness and Response at www.cdc.gov.
3. D. Dennis, T. Inglesby, D. Henderson, et al., "Tularemia as a biological weapon," *JAMA* 285 (June 6, 2001): 2763–73.
4. Associated Press, "Two cases of rabbit fever suspected on Martha's Vineyard," June 7, 2002.
5. Center for Civilian Biodefense Strategies/Agents/Tularemia at www.hopkinsbiodefense.org.
6. Ibid., Agents/Botulism.
7. Ibid., Agents/VHF.
8. Author interview with M. Galbraith.
9. T. Inglesby, T. O'Toole, D. Henderson, et al., "Anthrax as a biological weapon, 2002," *JAMA* 287 (May 1, 2002): 2236–52.
10. H. Kassenberg, R. Danila, P. Snikes, et al., "Human ingestion of bacillus anthracis-contaminated meat—Minnesota, August 2000," *JAMA* 284 (Oct. 4, 2000): 1644–45.
11. W. Broad and D. Grady, "Threats and responses: The victims; science slow to ponder the ills that linger in anthrax victims," *New York Times,* Sept. 16, 2002.
12. J. Warrick, "Anthrax specimens from army testing lab had no role in attacks," *Washington Post,* Jan. 22, 2002.
13. See www.cdc.gov. for information on the smallpox threat.
14. M. Enserink, "How devastating would a smallpox attack really be?" *Science* 296 (May 31, 2002): 1592–94.
15. Author interview with T. Monath.
16. R. Redfield, D. Wright, W. James, et al., "Disseminated vaccinia in a military recruit with human immunodeficiency virus (HIV) disease," *New England Journal of Medicine* 316 (Mar. 12, 1978): 673.
17. K. Alibek, *Biohazard* (New York: Random House, 1999).

18. S. Astbury, "Australia launches program to wipe out 170 million rabbits," *Deutsche Presse,* Oct. 11, 1996.

19. J. Cello, A. Paul, and E. Wimmer, "Chemical synthesis of poliovirus," *Science* 297 (Aug. 9, 2002): 1016–18.

20. B. Sellman, M. Mourez, and R. J. Collier, "Dominant-negative mutants of a toxin subunit: An approach to therapy of anthrax," *Science* 292 (Apr. 27, 2001): 695–97.

21. J. Cohen, "Blocking anthrax," *Science* 294 (Oct. 19, 2001): 500.

22. A. Revkin, "Can these boxes be locked against terror?" *New York Times,* Sept. 6, 2002.

23. CDC press release, "As Americans reflect on 9/11, HHS and CDC continue to aggressively prepare the nation for another terrorist attack," Sept. 2002 at www.cdc.gov.

24. Ibid.

25. Author interview with P. Della-Latta.

CHAPTER 3

1. J. McQuiston, "Two Suffolk County boys contract malaria at local Boy Scout Camp," *New York Times,* Sept. 1, 1999.

2. G. Ruiz, T. Rowlings, F. Dobbs, et al., "Global spread of microorganisms by ships," *Nature* 408 (Nov. 2, 2000): 49.

3. J. Robin, "Survivor's ordeal: debilitating virus leaves man facing a difficult recovery," *Newsday,* Sept. 14, 1999, and D. Kirschling, "Still loving the outdoors: First victim of West Nile is not afraid," *Newsday,* Aug. 6, 2000.

4. Author interview with M. Layton.

5. J. Koplan and S. Thacker, "Fifty years of epidemiology at the Centers for Disease Control and Prevention," *American Journal of Epidemiology* 154 (Dec. 1, 2001): 982–84.

6. J. Steinhauer and J. Miller, "In New York outbreak glimpse of gaps in biological defenses," *New York Times,* Oct. 11, 1999.

7. R. Craven and J. Roehrig, "West Nile virus," *JAMA* 286 (Aug. 8, 2001): 651–53.

8. T. Briese, X. Jia, C. Huang, et al., "Identification of a kunjin/West Nile–like flavivirus in brains of patients with New York encephalitis," *Lancet* 354 (Oct. 9, 1999): 1261–62.

9. D. Nash, F. Mostashari, A. Fine, et al., "The outbreak of West Nile virus infection in the New York City area in 1999," *New England Journal of Medicine* 344 (June 14, 2001): 1807–14.

10. "West Nile virus surveillance and control," *City Health Information* 20 (May 2001).

11. Author interview with J. Miller.

12. F. Mostashari, M. Bunning, P. Kitsutani, et al., "Epidemic West Nile encephalitis, New York, 1999," *Lancet* 358 (July 28, 2001): 261–64.

13. "WHO warns against 'jet-setting' mosquitoes carrying malaria," *Agence France Presse*, Aug. 25, 2000.

14. R. McLean, S. Ubico, D. Docherty, et al., "West Nile virus transmission and ecology in birds," *Annals of the New York Academy of Sciences* 951 (Dec. 2001): 54–57.

15. C. Chow, S. Montgomery, D. O'Leary, et al., "Provisional surveillance summary of the West Nile virus epidemic," *Morbidity and Mortality Weekly Reports* 51 (Dec. 2002): 1129–31.

16. "New arenavirus blamed for recent deaths in California," *Science* 289 (Aug. 11, 2000): 842–43.

CHAPTER 4

1. Bovine Spongiform Encephalopathy at www.who.int.

2. M. Enserink, "American's own prion disease?" *Science* 292 (June 1, 2001): 1641.

3. R. Imrie, "Deaths raise concerns chronic wasting disease may also affect humans," *Ottawa Citizen*, Aug. 2, 2002, and J. Davis, J. Kazmierczak, M. Wegner, et. al., "Fatal degenerative neurologic illnesses in men who participated in Wild Game Feasts—Wisconsin, 2002," *Morbidity and Mortality Weekly Reports* 52 (Mar. 2003): 125–127.

4. Author interview with A. Beyless.

5. S. Prusiner, "Prions," *Proceedings of the National Academy of Sciences* 95 (Nov. 1998): 13363–83.

6. R. Will, J. Ironside, M. Zeidler, et al., "A new variant of Creutzfeldt-Jakob in the UK," *Lancet* 347 (1996): 921–25.

7. Author interview with P. Monk.

8. C. Cookson, "Joint venture to detect infectious diseases in blood 'Mad Cow' disease," *Financial Times* (London), Apr. 9, 2002.

9. M. Balter, "Uncertainties plague projections of vCJD toll," *Science* 294 (Oct. 26, 2001): 770–71.

10. R. Imrie, "Deaths raise concerns," *Ottawa Citizen,* Aug. 2, 2002.

11. E. Belay, P. Gambetti, L. Schonberger, et al., "Creutzfeld-Jakob disease in unusually young patients who consumed venison," *Archives of Neurology* 58 (Oct. 2001): 1673–78.

12. D. Perlman, "Doctors study brain tissue of elk hunter," *San Francisco Chronicle,* Sept. 7, 2002.

13. Associated Press, "A look at chronic wasting disease in Wisconsin," Sept. 24, 2002.

14. B. Ripenhoff, "Guidelines created to increase safety of venison," *Milwaukee Journal Sentinel,* Sept. 8, 2002.

15. S. Blakeslee, "A marker for Mad Cow disease may be found in urine," *New York Times,* July 17, 2001.

16. M. Shareif, A. Green, J. Dick, et al., "Heightened intrathecal release of proinflammatory cytokines in Creutfeldt-Jakob disease," *Neurology* 52 (Apr. 12, 1999): 1289–91.

17. F. Heppner, C. Musahl, I. Arrighi, et al., "Prevention of scrapie pathogenesis by transgenic expression of anti-prion antibodies," *Science* 294 (Oct. 5, 2001): 178–82.

18. C. Korth, B. May, F. Cohen, and S. Prusiner, "Acridine and phenothiazine derivatives as pharmacotherapies for prion disease," *Proceedings of the National Academy of Sciences* 98 (Aug. 14, 2001): 9836–41.

19. S. Blakeslee, "Doctors test therapy for a brain malady," *New York Times,* Aug. 14, 2001.

20. P. Follette, "Prion disease treatment's early promise unravels," *Science* 299 (Jan. 10, 2003): 191–92.

21. D. Poretz, R. Williamson, K. Kaneko, et al., "Antibodies inhibit prion propagation and clear cell cultures of prion infectivity," *Nature* 412 (Aug. 16, 2001): 739–43.

CHAPTER 5

1. J. Warrick, "An outbreak waiting to happen: Beef inspection failures let in a deadly microbe," *Washington Post,* Apr. 9, 2001, and author interview with Connie Kriefall.

2. C. Wong. S. Jelacic, R. Habeeb, et al., "The risk of hemolytic-uremic syndrome after antibiotic treatment of *Escherichia coli* O157:H7 in-

fections," *New England Journal of Medicine* 342 (June 29, 2000): 1930–36.

3. N. Safdar, A. Said, R. Gagnon, and D. Maki, "Risk of hemolytic uremic syndrome after antibiotic treatment of *Escherichia coli* O157:H7 enteritis," *JAMA* 288 (Aug. 28, 2002): 996–1001.
4. Food-borne infections/technical information at www.cdc.gov; D. Satcher, "Food safety: A growing global health problem," *JAMA* 283 (Apr. 12, 2000): 1817.
5. See food safety inspection service (fsis) at USDA website www.usda.gov.
6. Author interview with R. Welch.
7. F. Blattner, G. Plunkett III, G. Bloch, et al., "The complete genome sequence of *Escherichia coli* K12," *Science* 277 (Sept. 5, 1997): 1453–62.
8. N. Perna, G. Plunkett III, V. Burland, et al., "Genome sequence of enterohaemorrhagic *Escherichia coli* O157:H7," *Nature* 409 (Jan. 25, 2001): 529–33.
9. E. Strauss, "Fighting bacterial fire with bacterial fire," *Science* 290 (Dec. 22, 2000): 2231–33.
10. P. Ray, D. Acheson, R. Chitrakar, et al., "The investigators of the hemolytic uremic syndrome Synborb multicenter clinical trial," *Journal of the American Society of Nephrology* 13 (March 2000): 699–707.
11. See STOP website at www.STOP-usa.org, or call 1–800–350–STOP.
12. Author interview with K. Taylor Mitchell.
13. *Supreme Beef Processors, Inc. v. United States Department of Agriculture,* No. 00–1108, United States Court of Appeals for the Fifth Circuit District, Dec. 6, 2001.
14. C. Drew and E. Becker, "Plants sanitation may have link to deadly bacteria," *New York Times,* Dec. 11, 2002.

CHAPTER 6

1. B. LeGere, "Bear Matthews grateful he's still alive," *Chicago Daily Herald,* July 22, 2000.
2. S. Palumbi, "Humans as the world's greatest evolutionary force," *Science* 293 (Sept. 7, 2001): 1786–90.
3. S. Tsiodras, H. Gold, G. Sakoulas, et al., "Linezolid resistance in a clinical isolate of *Staphylococcus aureus,*" *Lancet* 358 (July 21, 2000): 207–8.

4. Author interview with Stuart Levy.

5. Author interview with Keiichi Hiramatsu.

6. M. Kuroda, T. Okta, B. Yuzawa, et al., "Whole genome sequencing of a methicillin-resistant *Staphylococcus aureus*," *Lancet* 357 (Apr. 21, 2001): 1225–39.

7. D. Sievert, M. Boulton, G. Stoltman, et al., "*Staphylococcus aureus* resistant to vancomycin—United States 2002," *Morbidity and Mortality Weekly Reports* 51 (2002): 565–67.

8. J. Linder and R. Stafford, "Antibiotic treatment of adults with sore throats by community primary care physicians," *JAMA* 286 (Sept. 12, 2001): 1181–86.

9. S. Projan and P. Youngman, "Antimicrobials: New solutions badly needed," *Current Opinion in Microbiology* 5 (2002): 463–65.

10. As reported in the *London Daily Mail*, May 30, 2000.

11. J. Parienti, P. Thibon, R. Heller, et al., "Hand-rubbing with an acqueous alcoholic solution vs. traditional surgical hand-scrubbing and 30-day surgical site infection rates," *JAMA* 288 (Aug. 14, 2002): 722–27; see also "Hand hygiene in health care settings—2002," at www.cdc.gov.

12. D. White, S. Zhao, R. Sudler, et al., "The isolation of antibiotic resistant salmonella from retail ground meats," *New England Journal of Medicine* 345 (Oct. 18, 2001): 1147–54.

13. "An outbreak of multi-drug-resistant, quinolone-resistant *Salmonella enterica* serotype typhimurium DT 104," *New England Journal of Medicine* 341 (1999): 1420–25; see also D. Ferber, "Superbugs on the hoof?" *Science* 288 (May 5, 2000): 792–94.

14. S. Levy, "Antibacterial household products: Cause for concern," *Emerging Infectious Diseases* 7, suppl. 3 (June 2001): 512–15.

CHAPTER 7

1. See www.cdc.gov for information on influenza virus.

2. J. Lee, "Containing the Hong Kong flu outbreak," *Agricultural Research* (Dec. 1998) at www.ars.usda.gov.

3. Author interview with N. Cox.

4. B. Vastag, "Hong Kong flu still poses pandemic threat," *JAMA* 288 (Nov. 20, 2002): 2391–95.

5. "Summary of notifiable diseases, United States, 1998," *Morbidity and Mortality Weekly Reports* 47 (Dec. 31, 1999): 78–81.

6. M. Danovaro-Holliday, C. LeBaron, C. Allensworth, et al., "A large rubella outbreak with spread from the workplace to the community," *JAMA* 284 (Dec. 6, 2000): 2733–39.

7. P. Offit, "Addressing parents' concerns: Do multiple vaccines overwhelm or weaken the infant immune system?" *Pediatrics* 109 (Jan. 2002): 124–29.

8. Author interview with R. Strikas.

9. Author interview with L. Walsh.

10. Author interview with B. Fass-Offit.

11. J. Cohen, "U.S. vaccine supply falls seriously short," *Science* 295 (Mar. 15, 2002): 1998–2001.

12. Author interview with D. Ambrosino.

CHAPTER 8

1. Tuberculosis fact sheet no. 104 at www.who.org.

2. N. Schluger and W. Rom, "The host immune response to tuberculosis," *American Journal of Respiratory and Critical Care Medicine* 157 (1998): 679–91.

3. For various evasion techniques, see Z. Malik, G. Denning, and D. Kusner, "Inhibition of Ca2+ signaling by *Mycobacterium tuberculosis* is associated with reduced phagosome-lysosome fusion and increased survival within human macrophages," *Journal of Experimental Medicine* 191 (Jan. 17, 2000): 287–302; B. Li, H. Bassiri, M. Rossman, et al., "Involvement of the Fas/Fas ligand pathway in activation-induced cell death of mycobacteria-reactive human T cells," *Journal of Immunology* 161 (Nov. 1, 1998): 4983–91; D. Piddington, F. Fang, T. Laessig, et al., "Cu,Zn superoxide dismutase of *Mycobacterium tuberculosis* contributes to survival in activated macrophages that are generating an oxidative burst," *Infection and Immunity* 69 (Aug. 2001): 4980–87; and S. Redpath, P. Ghazal, N. Gascoigne, et al., "Highjacking and exploitation of IL-10 by intracellular pathogens," *Trends in Microbiology* 9 (Feb. 1, 2001): 86–92.

4. J. Donnelly, "Drug-resistant TB spreading, report says," *Boston Globe,* Mar. 24, 2000; J. Coleman, "Tuberculosis makes sharp comeback in Japan: Royals tested," *Associated Press* (Nov. 26, 2000).

5. J. Stephenson, "IOM report a blueprint for elimination of TB," *JAMA* 283 (June 7, 2000): 2776–77, or at www.iom.edu.

6. "California school becomes notorious for epidemic of TB," *New York Times,* July 18, 1994.

7. S. Cook, I. Blair, and M. Tyers, "Outbreak of tuberculosis associated with a church," *Communicable Diseases and Public Health* 3 (Sept. 2000): 181–83.
8. L. Garrett, *Betrayal of Trust,* Hyperion, 2000.
9. K. Brudney and J. Dobkin, "Resurgent tuberculosis in New York City," *American Review of Respiratory Diseases* 144 (Oct. 1991): 745–49.
10. Author interview with A. DeMaria.
11. Author interview with M. Malone.
12. Tuberculosis fact sheet no. 104 at www.who.org; K. Floyd, L. Blanc, M. Raviglione, and J.-W. Lee, "Resources required for global tuberculosis control," *Science* 295 (Mar. 15, 2002): 2040–42.
13. R. Gupta, J. Kim, M. Espinal, et al., "Responding to market failures in tuberculosis," *Science* 293 (Aug. 10, 2001): 1049–51.
14. J. McKinney, "In vivo veritas: The search for TB drugs goes live," *Nature Medicine* 6 (Dec. 2000): 1330–33.
15. R. Sharma, D. Saxena, A. Dwivedi, and A. Misra, "Inhalable microparticles containing drug combinations to target alveolar macrophages for treatment of pulmonary tuberculosis," *Pharmaceutical Research* 18 (Oct. 2001): 1405–10.
16. Author interview with H. Kornfeld.

CHAPTER 9

1. See www.cdc.gov. for information on hepatitis B.
2. C. Crabb, "Hard-won advances spark excitement about hepatitis C," *Science* 294 (Oct. 19, 2001): 506–7.
3. www.cdc.gov; J. Cohen, "The scientific challenge of hepatitis C," *Science* 285 (July 2, 1999): 26–27; also see G. Lauer and B. Walker, "Hepatitis C virus infection," *New England Journal of Medicine* 345 (July 5, 2001): 41–52.
4. A. Takaki, M. Wiese, G. Maertens, et al., "Cellular immune responses persist and humoral responses decrease two decades after recovery from a single source outbreak of hepatitis C," *Nature Medicine* 6 (May 2000): 578–82.
5. B. Meier, "Hepatitis cases may be linked to reuse of needles at clinic," *New York Times,* Oct. 17, 2002.
6. C. Frank, M. Mohamed, G. Strickland, et al., "The role of antischistosomal therapy in the spread of hepatitis C virus in Egypt." *Lancet* 355 (Mar. 11, 2000): 887–91.

7. Author interview with D. Nathan.

8. D. Mercer, D. Schiller, J. Elliot, et al., "Hepatitis C virus replication in mice with chimeric human livers," *Nature Medicine* 7 (Aug. 2001): 927–33.

9. M. Manns, J. McHutchison, S. Gordon, et al., "Peginterferon alfa-2b plus ribavirin compared with interferon 2b plus ribavirin for initial treatment of chronic hepatitis C," *Lancet* 358 (Sept. 22, 2001): 958–65.

10. R. Whitely, D. Kimberlin, B. Roizman, et al., "Herpes simplex viruses," *Clinical Infectious Diseases* 26 (1998): 541–55.

11. R. Whitley and J. Gnann, "Viral encephalitis: Familiar and emerging pathogens," *Lancet* 359 (Feb. 9, 2002): 507–14.

12. H. Burgert, "Subversion of the MHC class I antigen presentation pathway by adenoviruses and herpes simplex viruses," *Trends in Microbiology* 4 (Mar. 1996): 107–12.

13. C. Posavad, D. Koelle, and L. Corey, "Tipping the scales of herpes simplex virus reactivation: The important responses are local," *Nature Medicine* 4 (Apr. 1998): 381–82.

14. K. Eidson, W. Hobb, B. Manning, et al., "Expression of herpes simplex virus ICPO inhibits the induction of interferon-stimulated genes by viral infection," *Journal of Virology* 76 (Mar. 2000): 2180–90.

15. A. L. Hughes, "Origin and evolution of viral interleukin-10 and other DNA virulence genes with vertebrate homologs," *Journal of Molecular Evolution* 54 (Jan. 2002): 90–101.

16. J. Tanner and C. Alfieri, "Epstein-Barr virus induced Fas (CD95) in T cells and fas ligand in B cells leading to T cell apoptosis," *Blood* 94 (Nov. 15, 1999): 3439–47.

17. Author interview with H. Ploegh.

18. W. Johnson and R. Desrosiers, "Viral persistence: HIV's strategy of immune system evasion," *Annual Review of Medicine* 53 (2002): 499–518.

19. Author interview with D. Pillay.

20. A. Perelson, A. Neumann, M. Markowitz, et al., "HIV dynamics in vivo," *Science* (Mar. 15, 1996): 1582–86.

21. V. DiMartino, P. Rufat, N. Boyer, et al., "The influence of human immunodeficiency virus coinfection on chronic hepatitis C in injection drug users," *Hepatology* 34 (Dec. 2001): 1193–99.

22. T. Bingham, W. McFarland, D. Shehan, et al., "Unrecognized HIV infection, risk behaviors, and perception of risk among young black men who have sex with men—six U.S. cities, 1994–1998," *Morbidity and Mortality Weekly Reports* 51 (Sept. 2002): 733–36.

CHAPTER 10

1. Op cit. The Central Intelligence Agency Report.
2. T. S. Weiss, "Eat dirt: The hygiene hypothesis and allergic diseases," *New England Journal of Medicine* 347 (Sept. 19, 2002) 930–31.
3. D. Nathan, "Careers in translational clinical research: Historical perspectives, future challenges," *JAMA* 287 (May 8, 2002): 2424–27.
4. C. Holden, "Alliance launched to model *E. coli,*" *Science* 297 (Aug. 30, 2002): 1459–60.
5. N. Williams, T. Hirst, T. Nashar, et al., "Immune modulation by the cholera-like enterotoxins: From adjuvant to therapeutic," *Immunology Today* 20 (Feb. 1999): 95.
6. M. Brockman and D. Knipe, "Herpes simplex virus vectors elicit durable immune responses in the presence of preexisting host immunity," *Journal of Virology* 76 (Apr. 2002): 3678–87.
7. J. Keaney Jr. and J. Vita, "The value of inflammation for predicting unstable angina." *New England Journal of Medicine* 347 (July 4, 2002): 55–57.
8. L. Koutsky, K. Ault, C. Wheeler, et al., "A controlled trial of a human papillomavirus type 16 vaccine," *New England Journal of Medicine* 347 (Nov. 21, 2002): 1645–51.
9. D. Korn, R. Rich. H. Garrison, et al., "The NIH budget in the 'post-doubling' era," *Science* 296 (May 24, 2002): 1401–2.

EPILOGUE

1. L. Altman, "Flu season seems over, officials say," *New York Times,* March 8, 2004.
2. Avian flu at www.who.int/csr/disease/avian_influenza/en.
3. S. Gamblin et al., "The structure and receptor-binding properties of the 1918 influenza hemagglutinin," *Sciencexpress,* Feb. 6, 2004, and J. Stevens, A. Corper, C. Basler, et al., "Structure of the uncleaved human H1 hemagglutinin from the extinct 1918 influenza virus," *Sciencexpress,* Feb. 6, 2004.

4. USDA Statement "BSE Update—Friday January 30, 2004," at www.USDA.gov.

5. C. Casalone, G. Zanusso, P. Acutis, et al., "Identification of a second bovine amyloidotic spongiform encephalopathy: Molecular similarities with sporadic Creutzfeldt-Jakob diseases," *Proceedings of the National Academy of Sciences* 101 (March 2, 2004): 3065–70.

6. C. Llewelyn, P. Hewitt, R. Knight, et al., "Possible transmission of variant Creutzfeldt-Jakob disease by blood transfusion," *Lancet* 363 (Feb. 2004): 417–21.

7. Associated Press, "Seattle area family sues chain in mad cow claim," March 5, 2004.

8. "Update: Multistate outbreak of monkeypox—Illinois, Indiana, Kansas, Missouri, Ohio, and Wisconsin, 2003," *JAMA* 290 (July 23/30, 2003): 454–55.

9. K. Lutz, "Exotic meat imports feared for disease link, monkey meat seized at Logan last year," *Boston Globe,* July 17, 2003.

10. "WHO consensus document on the epidemiology of severe acute respiratory syndrome," at www.who.int/entity/csr/sars/en/WHOconsensus.pdf.

11. Chinese SARS Molecular Epidemiology Consortium, "Molecular evolution of the SARS coronavirus during the course of the SARS epidemic in China," at www.sciencexpress.org/29January2004.

12. B. Vastag, "Ebola vaccine tested in humans, monkeys," *JAMA* 291 (Feb. 4, 2004): 549–50.

13. L. Altman, "Pocket of opposition to vaccine threatens polio eradication," *New York Times,* Dec. 9, 2003; "Measles epidemic attributed to inadequate vaccination coverage—Campania, Italy," *JAMA* 290 (Dec. 3, 2003): 2792–3; R. Perez-Pena, "Refusal of vaccination cited in whooping cough outbreak," *New York Times,* Oct. 7, 2003.

14. M. Busch, L. Pietrelli, K. Sazama, et al., "Update: Detection of West Nile virus in blood donations—United States, 2003," *JAMA* 290 (Nov. 5, 2003): 2248–49. G. Peter, "Strengthening the supply of routinely recommended vaccines in the United States," *JAMA* 290 (Dec. 17, 2003): 3122.

ACKNOWLEDGMENTS

There are many rewards in writing a book. Chief among them, for us, was the interaction we enjoyed with smart sources, the insight we gained from victims and their families, and the help offered by talented publishing people.

We'd like to thank our wonderful agent, Jill Kneerim, who took Elinor's original idea about how successful microorganisms are becoming at outsmarting our immune defenses and suggested how to transform that into something of large interest to the reading public. Betsy Rapoport, our first editor at Crown, encouraged us to expound on the role that scientists and health professionals play in battling bacteria and viruses. When Betsy left Crown, Emily Loose added our book to her already full list. Emily accepted it as one of her own, and she worked tirelessly to give the book greater voice and make it more engaging. We also thank assistant editor Caroline Sincerbeaux, who helped us realize when we needed to be clearer.

This book really didn't come to life until we began talking with the families of victims who died at the hands of infectious diseases, and to patients who suffered terribly but survived. We would like to thank all of them for their willingness to bravely relive painful memories and divulge personal details so that readers can better know how to recognize the symptoms of infectious diseases and perhaps save their own lives or those of other people. In particular, we would like to recognize Jeannie Brown's mother, Audrey Trantham, her sister-in-law, Sherry Trantham, and friends Vivian

Howard and Cassandra Ledbetter; and Pamela Beyless's father and mother, Arthur and June Beyless, and friends Trudy Hewett and Cindy Hall. Among the survivors, we thank (in order of appearance) David Hose, Enrico Gabrielli (and his wife, Caterina), Julia Corazon, Suzie Nackiwanda, David Nathan, and Seth Rogovoy, as well as those—including John Tracy and Judy Moore—who shared stories that just couldn't fit into the book yet helped us fully appreciate the diabolical nature of infectious diseases.

Our research would fill linear yards of file space if we printed it all. Basic information came from a slew of biomedical sources, but took on meaning as we interviewed scores of researchers, doctors, health professionals, and patients. Many are cited by name in the chapters and notes, but we are especially grateful to the following people in the United States, United Kingdom, and Japan for talking with us and in some cases critiquing the text: Harvey Alter, Miriam Alter, Donna Ambrosino, Deborah Asnis, Paul Biedrzycki, Stephen Brossette, Calvin Cohen, Nancy Cox, Phyllis Della-Latta, Alfred DeMaria, Nancy Donley, Seth Foldy, Bonnie Fass-Offit, Neil Ferguson, Kathryn Freyfogle, Mark Galbraith, Keiichi Hiramatsu, Michael Keogh, Nick Komar, Hardy Kornfeld, Stuart Levy, Marci Layton, Michael Malone, James Miller, Michael Miller, Karen Taylor Mitchell, Tom Monath, Philip Monk and Gerry Bryant, David Nathan, Deenan Pillay, Hidde Ploegh, Peter Smith, Ray Strikas, Linda Walsh, and Rod Welch. Additional thanks go to bacterial geneticist Sue Fisher and physicians Andrew Bedford and Al DeMaria for reviewing large sections of the book. We take full responsibility for any errors that might have made it into print, and of course all opinions are our own.

We traveled across the United States for this book as well as crossing the oceans east and west. For their hospitality and logistical help in Britain we'd like to thank Janine and Lyndon Cole and Ben Manning, as well as Yoko Hatakeyama and the Tokyo American Center in Japan.

We certainly want to thank our own family members for their patience and support. Elinor is particularly grateful to her son, Ben, who slogged through early versions of the text and organized a contest among his friends to find a book title (sorry, none made it, but some were quite good). Mark would like to thank his wife, JoAnne, and his son, Nicholas, who both never grew tired of feeding Mark news reports on the latest outbreak or of asking, "So, is chapter X done yet?"

—E.L., M.F.

INDEX

ABOUT THE AUTHORS

ELINOR LEVY, PH.D.

is an immunologist and associate professor of microbiology at Boston University and the Boston Medical Center. Her research on the HIV virus and immune responses in healthy individuals has been presented in Africa, South America, and Europe, as well as in the United States. Levy has served as a grant application reviewer for the National Institutes of Health and for the MacArthur Foundation and is coauthor with Tom Monty of *The Ten Best Tools to Boost Your Immune System* (Houghton Mifflin, 1997).

MARK FISCHETTI

is a contributing editor to *Scientific American* and a veteran science writer. He also writes for *Technology Review* and has written for the *New York Times* and *Smithsonian.* He has also been coeditor of the quarterly *Scientific American Presents* and is coauthor with Tim Berners-Lee of *Weaving the Web* (HarperCollins, 1999).